Mathematical Modelling

Mathematical Modelling

D. Burghes
P. Galbraith
N. Price
A. Sherlock

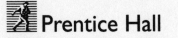 Prentice Hall

*London   New York   Toronto   Sydney   Tokyo   Singapore
Madrid   Mexico City   Munich*

First published (1996) by
Prentice Hall International (UK) Limited
Campus 400, Maylands Avenue
Hemel Hempstead
Hertfordshire, HP2 7EZ
A division of
Simon & Schuster International Group

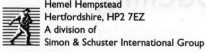

Printed and bound in Great Britain by
T J Press (Padstow) Ltd

Library of Congress Cataloging-in-Publication Data

Mathematical modelling / David Burghes . . . [et al.].
     p.   cm.
   Includes index.
   ISBN 0-13-290826-3
   1. Mathematical models.   I. Burghes, David N.
QA401.M3934   1996
658.4'0352–dc20   95–31832
                                                        CIP

British Library Cataloguing in Publication Data

A catalogue record for this book is available
from the British Library

ISBN 0-13-290826-3

1 2 3 4 5   00 99 98 97 96

Contents

Preface

In this resources text the important concept of mathematical modelling is introduced and developed through a number of case studies. These case studies are based on problems in the companies

British Steel

The Post Office

Esso

Tesco

We hope that this text will offer students taking A-level mathematics courses a useful extra set of resources to help them to gain confidence in applying mathematics to real world problems. The data section for each company provides real data and ideas for analysis and will be especially useful to students undertaking statistical project work.

Acknowlegements

The authors of this text are **David Burghes** with **Peter Galbraith**, **Nigel Price** and **Alan Sherlock**.

We are grateful for help and encouragement from the four companies, *British Steel, The Post Office*, *Esso* and *Tesco*, in the development of this text. Any errors, though, are the sole responsibility of the authors.

Camera-ready copy has been prepared by Liz Holland at

Centre for Innovation in Mathematics Teaching
School of Education
University of Exeter
Exeter EX1 2LU

CHAPTER ONE

Introduction

1.1 Nature of mathematics

Throughout history, mathematics has played a major role. It has been used for peaceful constructive help in solving problems and explaining phenomena and also for more aggressive purposes such as in the development of ever more sophisticated weapons.

Often the mathematics needed for solving problems has already been developed by pure mathematicians who study only particular aspects and concepts in mathematics, building on what has already been proved but with little thought to how it will eventually be used.

Other aspects of mathematics have been developed in response to real world problems – this has often been the case with mechanics. Solutions are needed to explain phenomena or predict results, and the mathematics has been developed specially to solve the problems.

In this text you will explore how mathematics, particularly topics central to A-level mathematics, can be used to help in solving real world problems through the development and exploration of mathematical models.

In this chapter, the main task is to see how mathematics can be used to

- explain
- predict
- make decisions.

You will need to work through this text; if there is time, try the examples before you work through the solutions given.

You should also try the Activities and the Exercises; note that the 'stop and think' points are printed in this typeface.

1.2 Applying mathematics

Bode's Law

In 1772, the German astronomer *Johann Bode* (1747–1826) investigated the pattern formed by the distances of planets from the sun.

At that time only six planets were known and the pattern Bode devised is shown below. The distances are measured on a scale that equates 10 units to the sun/earth distance.

Planet	Actual distance	Bode's pattern
Mercury	4	0 + 4 = 4
Venus	7	3 + 4 = 7
Earth	10	6 + 4 = 10
Mars	15	12 + 4 = 16
–	–	–
Jupiter	52	48 + 4 = 52
Saturn	96	96 + 4 = 100

What do you think is the missing entry?

There are also planets further out than Saturn.

Find the next two numbers in Bode's pattern.

In fact, the data continues as shown here.

Planet	Actual distance
Uranus	192
Neptune	301
Pluto	395

Can you give an explanation of how Bode's Law can be adapted for this extra data?

Wind chill

When the temperature drops near zero, it is usual for weather forecasters to give both the expected air temperature and the *wind chill* temperature – this is the temperature actually felt by someone, which depends on the wind speed and air temperature. So, for example, the wind chill temperature for an actual temperature of 0°C and wind speed of 10 mph is given approximately by –5.5°C.

For $v > 5$ mph, the wind chill temperature is given by

$$T = 33 + \left(0.45 + 0.29\sqrt{v} - 0.02v\right)(t - 33)$$

where t°C is the air temperature and v mph the wind speed. This formula was devised by American scientists during the Second World War, and is based on experimental evidence.

Example

Find the wind chill temperature when

(a) $t = 2°C$, $v = 20$ mph

(b) $t = -10°C$, $v = 5$ mph

(c) $t = 0°C$, $v = 40$ mph

Solution

(a) When $t = 2$, $v = 20$,

$$T = 33 + \left(0.45 + 0.29\sqrt{20} - 0.02 \times 20\right)(-31)$$
$$\approx -8.8°C$$

(b) When $t = -10$, $v = 5$

$$T = 33 + \left(0.45 + 0.29\sqrt{5} - 0.02 \times 5\right)(-43)$$
$$\approx -9.9°C$$

(c) When $t = 0°C$, $v = 40$ mph,

$$T = 33 + \left(0.45 + 0.29\sqrt{40} - 0.02 \times 40\right)(-33)$$
$$\approx -16.0°C$$

What is the significance of a wind speed of about 5 mph?

Heptathlon

The heptathlon is a competition for female athletes who take part in seven separate events (usually spread over a two-day period).

For each event there is a points scoring system, based on the idea that a good competitor will score 1000 points in each event. For example, the points scoring system for the 800 metres running event is

$$\boxed{P = 0.11193(254 - m)^{1.88}}$$

where m is the time taken in seconds for the athlete to run 800 metres.

Example

For $m = 124.2,$

$$P = 0.11193(254 - 124.2)^{1.88}$$

$$\approx 1051.72$$

$\Rightarrow$ $P = 1051$

(Scores are always rounded down to the nearest whole number.)

Now, to score 1000 points requires a time of m seconds where

$$1000 = 0.11193(254 - m)^{1.88}$$

$$\Rightarrow (254 - m)^{1.88} = 8934.15$$

$$\Rightarrow \quad 254 - m = (8934.15)^{\frac{1}{1.88}}$$

$$\Rightarrow \qquad m = 254 - 126.364$$

giving $m = 127.64$

For all track events a points scoring system of the form

$$P = a(b - m)^c$$

is used, with suitable constraints for a, b and c.

Suggest an appropriate form for the formula for a points system for track events in the heptathlon.

Simple pendulum

The great Italian scientist *Galileo* (1564–1642) was the first to make important discoveries about the behaviour of swinging weights. These discoveries led to the development of pendulum clocks.

Activity 1.1   Period of a pendulum swing

Attach a weight at one end of a light string, the other end being fixed. Let the pendulum swing freely through a small angle in a vertical plane and, for various lengths of pendulum, l, in metres, find the corresponding times in seconds of one complete oscillation (known as the *period*). It is more accurate to time, say, five oscillations and then divide the total time by five. On a graph, plot the period, T, against the square root of the pendulum length, $\sqrt{l}$. What do you notice?

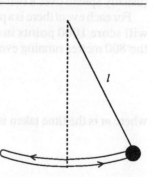

In fact, the two quantities are related approximately by the formula

$$T = 2.006\sqrt{l}$$

Example
What pendulum length gives a periodic time of one second?

Solution
If $T = 1$, then

$$1 = 2.006\sqrt{l}$$
$$\Rightarrow \sqrt{l} = \frac{1}{2.006} = 0.4985$$
$$\Rightarrow l \approx 0.25 \text{ m}$$

Activity 1.2

Construct a simple pendulum with $l = 0.25$ m and check its periodic time.

Exercise 1A

1.   Use the wind chill temperature formula to find its value where

      (a) $t = 0°C$, $v = 20$ mph

      (b) $t = 5°C$, $v = 20$ mph

      (c) $t = -5°C$, $v = 20$ mph

      Plot a graph of wind chill temperature against air temperature, t, for $v = 20$ mph.
      Use your graph to estimate the wind chill temperature when
              $t = 10°C$ and $v = 20$ mph.

2.   The points scoring system for the high jump event in the heptathlon is given by
              $$P = a(m - b)^c$$
      where $a = 1.84523$, $b = 75.0$, $c = 1.348$ and m is the height jumped, in
      centimetres. Find the points scored for a jump of 183 cm and determine the
      height required to score 1000 points.

3.   An algorithm for determining the date of Easter Sunday is given at the top of the
      next page. Use it to find the date of Easter Sunday next year, which is given by
      the pth day of the nth month.

Step	Number	Divide by	Answer (if needed)	Remainder
1	$x = $ year	100	$b = $	$c = $
2	$5b + c$	19	–	$a = $
3	$3(b + 25)$	4	$r = $	$s = $
4	$8(b + 11)$	25	$t = $	–
5	$19a + r - t$	30	–	$h = $
6	$a + 11h$	319	$g = $	–
7	$60(5 - s) + c$	4	$j = $	$k = $
8	$2j - k - h + g$	7	–	$m = $
9	$h - g + m + 110$	30	$n = $	$q = $
10	$q + 5 - n$	32	–	$p = $

Where a dash is shown, it means that the number is not needed.

4. Write a computer program to determine the date of Easter Sunday for the next 100 years. Illustrate the data using a histogram.

1.3 Mathematical modelling

Mathematics can be a very powerful tool in solving practical problems. This usually requires the development of a mathematical model to describe the real world situation. An example of this is given below.

Reading age formula

Educationalists need to be able to assess the minimum reading age of certain books so that the books can be appropriately catalogued, particularly for use with young children.

You are probably aware that, for example, it is much easier and quicker to read one of the tabloid newspapers (e.g. *The Sun*) than one of the quality broadsheets (e.g. *The Guardian*).

What factors influence the reading age of a book, newspaper or pamphlet?

There have been many attempts at designing a formula for finding the reading age of a text. One example is known as the *FOG formula*. This is given by

$$R = \frac{2}{5}\left(\frac{A}{n} + \frac{100L}{A}\right)$$

where the variables are defined for a sample passage of text by

A = number of words

n = number of sentences

L = number of words containing three or more syllables
(excluding '-ing' and '-ed' endings).

Activity 1.3

Find four or five books of different reading difficulty. First estimate the minimum reading ages for each of these, then use the FOG formula to compare the two sets of data.

Of course, the whole idea of giving a reading age for a particular book is perhaps rather dubious. Nevertheless, the problem is a real one and teachers and publishers do need to know the appropriate order for their reading books.

The example above illustrates the idea of a mathematical model; that is, a mathematical description of a problem. The mathematical model in the problem is essentially given by the equation for the FOG formula.

The translation of the problem from the real world to the mathematical world can be summarized in the diagram below.

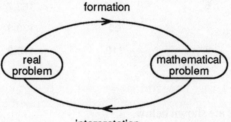

The model is formed from the real problem by making various assumptions, whilst the solution to the mathematical problem must be interpreted back in terms of the real problem. This will be illustrated in the next example.

Handicapping weightlifters

In weightlifting there are nine official bodyweight classes. For some competitions it is important to be able to compare lifts made by competitors in different classes. This means that some form of handicapping must be used.

There are a number of models that have been used to provide a form of handicapping. For example, if

L = lift (in kg), W = competitor's weight (in kg), L' = handicapped lift

then two possible solutions are

(a) $L' = L - W$

(b) $L' = L/(W - 35)^{\ddagger}$

The first method was used for some time in a television programme (*TV Superstars*) in which competitors of different weights competed against each other in a number of sports events. The second method, called the O'Carroll formula, is used in more serious competitions in order to find an overall winner.

Example

The best lifts (for the 'snatch' lift) for the competitors are given in the next table, together with their bodyweights. Use the two methods to find an overall winner.

Competitor	Weight (in kg)	Lift (snatch) (in kg)
1	52	105.1
2	56	117.7
3	60	125.2
4	67.5	135.2
5	75	145.2
6	82.5	162.7
7	90	170.3
8	110	175.3

Solution

The handicapped lifts are shown below.

Competitor	W	L	$L - W$	$L/(W - 35)^{\ddagger}$
1	52	105.1	53.1	40.9
2	56	117.7	61.7	42.7
3	60	125.2	65.2	42.8
4	67.5	135.2	67.7	42.4
5	75	145.2	70.2	42.5
6	82.5	162.7	80.2	44.9 ←
7	90	170.3	80.3 ←	44.8
8	110	175.3	65.3	41.6

For the first method, the winner is competitor number 7, but the second method makes competitor number 6 the winner.

Dicey game

For the school fete you have to organize a game of chance in which contestants pay a 10p entry fee to roll three dice. The total score is recorded and there are money prizes for high scores. The problem is to decide whether these prizes will give sufficient incentive for people to enter the game, and sufficiently low for the school to make a profit!

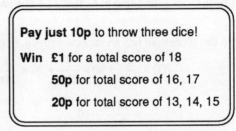

Pay just 10p to throw three dice!

Win £1 for a total score of 18

    50p for total score of 16, 17

    20p for total score of 13, 14, 15

The model to use here is straightforward – you associate a probability of $\frac{1}{6}$ with obtaining any number 1, 2, ..., 6 with one roll of a dice.

The mathematical problem is to determine the expected winnings/losses for a contestant. To find the expected winnings you must first find the probabilities of obtaining a total score of 18, 17, ..., 13 since these are the scores which win prizes.

What is the probability of obtaining 18?

Activity 1.4

Complete the table below, which gives the probabilities of a total score of 3, 4, ..., 18.

3	4	5	6	7	8	9	10	11	12	13	14	15	16	17	18
$\frac{1}{216}$	"	"	"	"	"	"	"	"	"	"	"	"	$\frac{6}{216}$	$\frac{3}{216}$	$\frac{1}{216}$

(Check that your sum of the probabilities is 1.)

With these values for the probabilities you can now determine the expected winnings/ losses for a contestant.

The winnings will be

$$£1 \text{ with a probability of } \frac{1}{216}$$

$$50p \text{ with a probability of } \frac{9}{216}$$

$$20p \text{ with a probability of } \frac{46}{216}$$

Check from your table in Activity 1.4 the values $\dfrac{9}{216}$ and $\dfrac{46}{216}$ used above.

So the contestant has an expected profit, in pence, of

$$100 \times \frac{1}{216} + 50 \times \frac{9}{216} + 20 \times \frac{46}{216} - 10$$

$$= \frac{1470}{216} - 10$$

$$= \frac{-690}{216}$$

$$\approx -3p \qquad \text{(a loss of 3p)}$$

Since, on average, each contestant will lose 3p, if 100 people play the game at the fete, the school will expect to make a profit of about

$$100 \times 3p = £3$$

which will not be of much help to school funds!

It is not really the model that is in need of improvement; the money prizes on offer should be changed.

Activity 1.5

Suggest new prizes so that, on average, a contestant will lose about 5p per game.

Whilst mathematics is a precise science, applications to real problems require both an understanding of the problem and an appreciation that, whilst mathematics can provide answers and give precise explanations based on particular assumptions and models, it cannot always solve the real problem.

Mathematics can help in the designing of multi-stage rockets that work but it cannot necessarily help to solve the problem of world peace. Often mathematical analysis can help in making the best decisions and, for example, the success of mathematical modelling is shown by the fact that man has walked on the moon. You should, though, be aware that most problems in real life are more complicated than a single equation or formula!

Activity 1.6

According to the *Guinness Book of Records* (1994), the tallest man of whom there is irrefutable evidence was *Robert Wadlow*, born on 22 February 1918 in Illinois (USA). The following table gives his height at various ages.

Age (in years)	Height	
5	5'4"	163 cm
9	6'2½"	189 cm
11	6'7"	210 cm
13	7'1¾"	218 cm
15	7'8"	234 cm
17	8'0½"	245 cm
19	8'5½"	258 cm
21	8'8¼"	265 cm

(Robert Wadlow died, at the age of 22, on 15 July 1940.)

Investigate whether Wadlow's growth could reasonably be described as linear.

Activity 1.7   Water flow

Take a plastic bottle and make a small hole near the bottom.
Fix a height scale on the bottle as shown in the diagram.

Cover the hole and fill the bottle with water up to the level of the top of the height scale.

Uncover the hole and measure the height (h) of the water surface above the hole against the time (t) since the hole was uncovered. Repeat this a number of times and find average values of t for, say, $h = 10, 9, 8, ..., 2, 1$ cm.

Does the data give support to either of the following models?

(a) $h = Ae^{-kt}$

(b) $h = at + b$

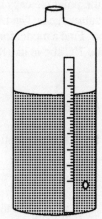

Activity 1.8   The greenhouse effect

The burning of fossil fuels such as coal and oil adds carbon dioxide to the atmosphere around the earth. This may be partly removed by biological reactions, but the concentration of carbon dioxide is gradually increasing. This increase leads to a rise in the average temperature of the earth. The following table shows this temperature rise over the 100-year period up to 1980.

Year	Temperature rise of the earth above the 1860 figure (°C)
1880	0.01
1896	0.02
1900	0.03
1910	0.04
1920	0.06
1930	0.08
1940	0.10
1950	0.13
1960	0.18
1970	0.24
1980	0.32

If the average temperature of the earth were to rise about another 6°C from the 1980 value, there would be a dramatic effect on the polar ice, caps, winter temperature, etc. As the polar ice caps melt, there could be massive floods and large areas of land could be submerged. The UK would disappear except for the tops of the mountains!

Find a model for the above data and use it to predict when the earth's temperature will be 7°C above its 1860 value. [Hint: plot a graph of log (temperature rise) against year.]

Exercise 1B

1. A mathematical model for the reading age of a text is given by

$$R = 25 - \frac{N}{10}$$

where N is the average number of one-syllable words in a passage of 150 words. Use this model to find the reading age of a number of books. Compare the results with those found in the case outlined in Section 1.3.

2. Use the handicapping model

$$L' = L / W^{\frac{1}{3}}$$

to find the winner of the competition described in the Example on page 8.

3. Blacksmiths make horseshoes by taking straight strips of iron and bending them into the usual horseshoe shape. To find what length of strip of iron is required, the blacksmith measures the width, W inches, of the shoe and uses a formula of the form

$$L = aW + b$$

to find the required strip length, L inches. Use the following data to find estimates for a and b.

Width W (inches)	Length L (inches)
6.50	12.00
5.75	13.50

4. A basic problem faced by some industries is how best to cut up lengths of tube which are supplied with fixed dimensions, in order to meet the required demand, with as little wastage as possible.

A simple example of this basic problem is to determine how many lengths of tube 10 m long are required to meet the following order and how they should be cut.

    60 lengths, each of 3 metres

    49 lengths, each of 4 metres

    12 lengths, each of 7 metres

If the basic lengths of tube are supplied in 12 metre lengths, what is the optimum answer?

A similar problem is that of sheets of metal which are supplied with fixed dimensions 10 m × 10 m. The following order is to be satisfied:

    60 sheets, each 3 m × 1 m

    49 sheets, each 4 m × 2 m

    12 sheets, each 7 m × 5 m

How should the sheets be cut?

What happens if the sheets are 12 m × 5 m?

5. The following table gives the average distance of the planets from the sun and the ratio of these distances to the earth's distance from the sun. Ignoring Mercury, assign a number to each planet in the following way:

    for   Venus choose $n = 0$

    Earth choose $n = 1$

    Pluto choose $n = 8$

Find a formula connecting R/R_e and n.

What value of n should be given to Mercury so that it also fits the model?

Planet	Distance from the sun (millions of km)	Ratio R/R_e where R_e is the earth's distance from the sun
Mercury	57.9	0.39
Venus	108.2	0.72
Earth	149.6	1.00
Mars	227.9	1.52
Asteroids	433.8	2.90
Jupiter	778.3	5.20
Saturn	1427.0	9.54
Uranus	2870.0	19.20
Neptune	4497.0	30.10
Pluto	5907.0	39.50

CHAPTER TWO

British Steel

2.1 Tin can design

Tin cans come in various shapes and sizes, but what factors influence their design?
In particular, is minimizing the area of tin used to make a can an important factor?

Suppose that a manufacturer wishes to enclose a fixed
volume V, using a cylindrical can, as shown. The height of the
cylinder is denoted by h and radius of the circular can-section
is r.

A net for the cylinder is shown below.

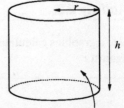

volume V

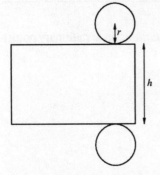

What is the length of the rectangle?

Since the circumference of the circular cross section is
equal to the length of the rectangle, it must be $2\pi r$, and
the total *surface area*, S, is given by

$$S = \pi r^2 + 2\pi rh + \pi r^2$$
$$= 2\pi r^2 + 2\pi rh$$

S is a function of two variables, r and h, but these are related by the volume formula
$$V = \pi r^2 h$$

Activity 2.1

Write S as a function of r by eliminating h.

Since $h = \dfrac{V}{\pi r^2}$, then

$$S = 2\pi r^2 + 2\pi r \left(\dfrac{V}{\pi r^2} \right)$$

$$\boxed{S = 2\pi r^2 + \dfrac{2V}{r}}$$

What does S tend to as $r \to \infty$?

What does S tend to as $r \to 0$?

Putting all these facts together results in a graph of the
form shown.   There will be a value $r = r_0$ which
minimizes the surface area.

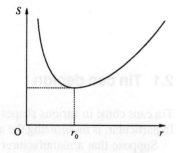

Activity 2.2

Use a graphics calculator, with say $V = 500\,\text{ml}$, to confirm the shape of the graph S
against r.

One method of finding this minimum value is to use *differentiation*, since stationary points
of the function S occur when

$$\dfrac{\mathrm{d}S}{\mathrm{d}r} = 0$$

This gives

$$\dfrac{\mathrm{d}S}{\mathrm{d}r} = 4\pi r - \dfrac{2V}{r^2}$$

$$= 0$$

when

$$4\pi r = \dfrac{2V}{r^2}$$

$$\Rightarrow \quad V = 2\pi r^3 \quad \Rightarrow \quad r_0 = \left(\dfrac{V}{2\pi} \right)^{\frac{1}{3}}$$

To find the optimum ratio of h to r, note that

$$V = \pi r_0^2 h = 2\pi r_0^3$$

$$\Rightarrow \quad h = 2r_0$$

or, alternatively

> height = diameter

How do we know that this stationary point is in fact a minimum?

One way to check that this is a *minimum* is to find the second differential

$$\frac{d^2S}{dr^2} = 4\pi + \frac{4V}{r^3}$$

and, at $r = r_0$, $\frac{d^2S}{dr^2} > 0$. This shows that there is a minimum of S at $r = r_0$.

Activity 2.3

For various tins, measure their height and diameter. Do your results support the argument that minimizing surface area is a key factor in the design of the tins?

Most tins do not seem to support the 'minimizing surface area' result.

Can you suggest important factors that were neglected in the analysis above?

It appears as if aesthetic appeal is of more importance in the design than minimum surface area. One case where aesthetic appeal is of little importance though, is that of commercial tins.

Activity 2.4

Repeat Activity 2.3 for commercial tins. Does the mathematical analysis hold for these tins?

Exercise 2A

1. What is the three-dimensional shape that, for a given volume, minimizes the surface area?

2. Repeat the problem of minimizing surface area for closed hexagonal cylinders (as shown) to enclose a fixed volume.

3. Find the ratio of height to diameter for a open-topped cylindrical container (as shown) which minimizes the surface area to enclose a fixed volume V.

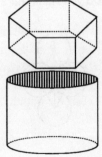

2.2 Drinks cans

Drinks cans are made by stamping out circular discs from a sheet of tin. The discs are bent up to make a shallow cup and then stretched to make a can. The tin gets thinner and harder during the process. The top is made separately and fixed on after the can has been made. The cans are usually stamped out from a sheet of tin as shown in the diagram. Given that the sheet is 2 m × 1 m, and that the radius of the circle stamped out is 10 cm, what wastage of tin is there and is there a more effective method?

Using the configuration shown opposite you can obtain $10 \times 5 = 50$ such cans from one sheet of metal. Each one gives the same amount of wastage, namely

$$\text{area of square} \ - \ \text{area of circle}$$
$$= 20 \times 20 - \pi \times 10 \times 10 .$$

So

$$\% \text{ wastage} = \frac{(400 - 100\pi)}{400} \times 100$$
$$= \left(1 - \frac{\pi}{4}\right)100$$
$$\approx 21\%$$

Activity 2.5

Assuming that the wastage can be recycled, and that can tops are stamped out in a similar way, what will be the percentage change in thickness of the tin in order to provide one top from the wastage in making one can?

Can you suggest other configurations for stamping out the cans?

Another possible way is shown opposite.

Do you think there is less waste using this configuration?

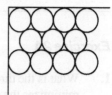

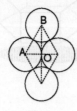

Using the dimension previously given

$$\cos O\hat{A}B = \frac{OA}{AB} = \frac{10}{10 + 10} = \frac{1}{2}$$

Hence $O\hat{A}B = 60°$ and the length OB is given by

$$OB = 20\sin 60° = 20 \times \frac{\sqrt{3}}{2} = 10\sqrt{3} \ \text{ (or by Pythagoras)}.$$

Activity 2.6

Using a rectangular sheet 2 m × 1 m (as before) and marking first a row of circles along the 2 m length, how many circles can be cut from the sheet using this configuration? Evaluate the total waste, and the percentage waste for the complete sheet.

Suggest reasons why this method is not used in practice.

Exercise 2B

1. What would be the percentage of waste using the second configuration if the sheet was very large (i.e. assume it to be infinite)?

2. Suppose the wastage in using the first design can be recycled to produce further cans. How many would be produced on the next cycle?

3. Investigate other two-dimensional shapes that need to be cut out in bulk from sheets of metal, tin or steel.

4. Investigate three-dimensional packing problems; for example, how can tennis balls be efficiently packed?

2.3 Wavy edges

Due to problems in the manufacture of tinplate coils, the edge of the strip can be slightly longer than the centre. This causes a 'wave' on the wall of the coil but can be rectified by differentially stretching the strip to make the edges flat.

As an example, suppose that the wave when laid flat on a table is found to have a height of 4 mm and a length of 250 mm as shown.

The percentage *elongation* needs to be determined in order to eliminate this wave condition. To achieve this calculation, it is assumed that the wave is an arc of a circle, so that

    percentage elongation = % difference between the chord and arc length

as illustrated below.

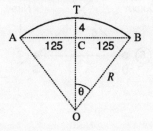

If the radius of the circle is assured to be R mm, then

$$OC = R - 4 .$$

How can you determine the length of R?

One method is to apply Pythagoras' theorem to triangle OBC; this gives

$$OB^2 = BC^2 + CO^2$$

$$\Rightarrow \quad R^2 = 125^2 + (R-4)^2$$

$$= 125^2 + R^2 - 8R + 16$$

$$\Rightarrow \quad 8R = 125^2 + 16$$

$$\Rightarrow \quad R = \frac{15641}{8} \approx 1955 \text{ mm}$$

Activity 2.7

Find the value of the angle θ and hence determine the arc length AB, and the percentage difference between this length and the chord AB.

Exercise 2C

1. Find the percentage elongation for a wave of height 5 mm and length 500 mm.

2. Generalize the results to find the percentage elongation for a wave of height h mm and length l mm.

2.4 Transportation

The cost of transportation is a key factor for many industries. A particular example is that of *British Steel's* stainless steel manufacturing plants. They require the raw material, ferrous oxide, which is mined and brought from Australia, South Africa or Japan. The actual cost of the raw material varies in each country, as does the cost of shipment to the UK. The problem faced by British Steel is to decide what quantities to bring from the available sources.

As a simpler example, consider two suppliers of a particular raw material. They can output 5 and 6 units per day respectively. The material is required at two factories, P and Q, which need 4 and 7 units per day. The cost of transporting one unit from the supplier to the factory is shown by the number on each arc in the diagram opposite.

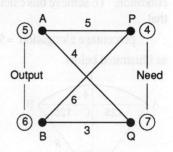

Activity 2.8

Find a way of supplying the factories, and calculate the total costs. Have you obtained the *minimum* cost solution? If not, find it.

With just two suppliers and two factories it is easy to find the optimum solution, but the problem becomes increasingly difficult to solve when we have more suppliers and factories.

For example, consider *three* suppliers and *two* factories as shown.

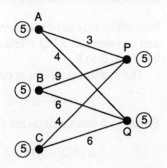

Activity 2.9

Find the minimum cost solution for the problem above.

An even more complicated example is shown opposite, in which there are *three* suppliers and *four* factories.

Activity 2.10

Find the minimum cost solution for this problem.

By now it is not quite so easy to see whether you have in fact obtained the *optimum* solution. Returning to the first example, suppose you have so far obtained the solution as shown; actual allocations are shown in squares.

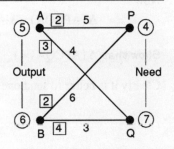

The total cost is given by

$$(2 \times 5 + 3 \times 4) + (2 \times 6 + 4 \times 3)$$
$$= 22 + 24 = 46$$

Is this the minimum cost solution?

Improvements can be made by sending more units from B to Q. For example, if one further unit was sent along BQ, one less would be sent from A to Q, one more from A to P and one less from B to P.

The increment in cost saving would be

$$3 - 4 + 5 - 6 = -2$$

So for each extra unit on the cycle B Q A P B the saving is 2, i.e.

$$\Delta (BQAPB) = -2$$

How many more units can be saved on this cycle?

Since the output of B is just 6 units, only two further units can flow in this way, giving a total saving of 4, and minimum cost of 42.

Check this answer with your answer to Activity 2.8.

This technique can be applied to more complex problems. With the allocation shown (in squares) the total cost is given by

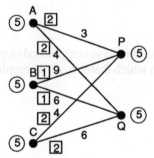

$$(2 \times 3 + 2 \times 4) + (1 \times 9 + 1 \times 6) + (2 \times 4 + 2 \times 6)$$
$$= 14 + 15 + 20$$
$$= 49$$

Sending one further unit from A to P might reduce the cost.
Possible 'cycles' are

$$A P B Q A$$
$$A P C Q A$$

Now

$$\Delta (APBQA) = 3 - 9 + 6 - 4 = -4$$

Show that $\Delta (APCQA) = 1$

Clearly it is better to send one further unit on the cycle A P B Q A.

Activity 2.11

What other cycles exist that will reduce the total cost? Use this method to find the minimum total cost.

For the minimum total cost, no cycles with negative increases in cost exist for which units can flow.

Exercise 2D

1. Use the 'cycle' method to check your solution to Activity 2.10. Can it be improved?

2. For each of the problems illustrated below find the minimum cost solution.

(a)

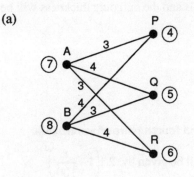

(b)

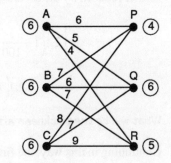

(c)

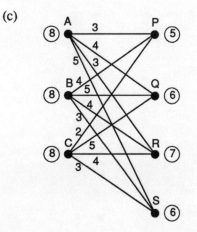

2.5 Reduction mill

The diagram opposite represents a
cold reduction mill.

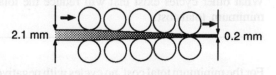

   It consists of five pairs of
workrolls. The incoming steel is of
thickness 2.1 mm, whilst the mill is
producing strip of thickness 0.2 mm.

If the percentage reduction is constant on each pair of workrolls, find that reduction. Also,
if the incoming speed is 1 m s^{-1}, what is the outgoing speed?

   This example uses mathematics similar to that of compound interest, except that it is
equivalent to negative interest. At the first pair of workrolls, the actual reduction is given
by

$$2.1 \times \left(1 - \frac{r}{100}\right)$$

if r per cent is the reduction produced by each pair of workrolls. This new thickness is
further reduced by the second pair of workrolls and the outgoing thickness will be

$$\left[2.1\left(1 - \frac{r}{100}\right)\right]\left(1 - \frac{r}{100}\right)$$

$$= 2.1\left(1 - \frac{r}{100}\right)^2$$

What will be the thickness after the third and fourth pairs of workrolls?

Continuing in this way, the final thickness will be given by $2.1\left(1 - \frac{r}{100}\right)^5$.

But this must equal 0.2, so

$$2.1\left(1 - \frac{r}{100}\right)^5 = 0.2$$

$$\Rightarrow \quad \left(1 - \frac{r}{100}\right)^5 = \frac{0.2}{2.1} = 0.095238$$

$$\Rightarrow \quad \left(1 - \frac{r}{100}\right) = (0.095238)^{\frac{1}{5}}$$

$$= 0.6248$$

$$\Rightarrow \quad \frac{r}{100} = 1 - 0.6248$$

$$= 0.3752$$

$$\Rightarrow \quad r = 37.52\%$$

Activity 2.12

If each pair of workrolls can, at most, provide 25 per cent reduction in thickness, how many pairs of workrolls are needed to reduce the thickness from 2.1 mm to 0.2 mm?

Solve the problem in general when each pair can reduce by p per cent.

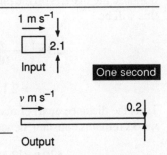

The exit speed can be determined from volume conservation – no steel is lost. So, in one second, the volume of steel being introduced into the mill is

$1 \times 2.1 \times$ width.

Assuming an output speed of v m s^{-1} (and the same width), then this must equal

$v \times 0.2 \times$ width.

Hence

$$2.1 = v \times 0.2$$

$$\Rightarrow \quad v = \frac{2.1}{0.2}$$

$$v = 10.5 \text{ m s}^{-1} \quad \text{(about 25 mph)} \quad \text{(a considerable increase)}.$$

Activity 2.13

The output speed is crucial for the next stage in the cycle of production, so it must be carefully controlled. If each pair of workrolls reduces the width by 30 per cent, find the width and output speed after 1, 2, ..., 10 pairs of workrolls for an input speed of 1 m s^{-1}.

Given the input speed and the input thickness and output thickness, does the number of pairs of workrolls and their reduction factor change the output speed?

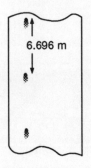

6.696 m

Returning to the first problem in which there are five pairs of workrolls with an input thickness of 2.1 mm and output thickness of 0.2 mm, a mark is noticed on the finished strip.

If the mark repeats itself every 6.696 metres and is attributable to a defect on one of the pairs of rolls, which pair is producing it? (The diameters of all the rolls are 520 mm.)

Suppose the mark is on the nth pair of workrolls; the thickness of the strip coming out of this will be

$$2.1 \times \left(1 - \frac{r}{100}\right)^n$$

$$= 2.1 \times (0.6248)^n.$$

Then the distance apart of marks would be $520\,\pi$ mm. But the volume of steel between marks remains constant, hence

$$520\pi \times \left[2.1 \times (0.6248)^n\right] \times \text{width}$$

$$= L \times 0.2 \times \text{width}$$

when $L = 6.696$ m $= 6696$ mm, the distances between marks on the final output strip.

Activity 2.14

Solve the equation above for n.

The equation can be written as

$$(0.6248)^n = \frac{6696 \times 0.2}{520\pi \times 2.1}$$

i.e. $(0.6248)^n = 0.3904$

How do you solve this equation for n?

A formal method is to take natural logarithms on both sides to give

$$n \ln(0.6248) = \ln(0.3904)$$

$$\Rightarrow \qquad n = \frac{\ln(0.3904)}{\ln(0.6248)}$$

$$= 2$$

So the fault is with the second pair of workrolls.

Exercise 2E

1. Find a general formula to give the output thickness and speed in terms of:

n, the number of pairs of workrolls;

r, the percentage reduction given by each pair of workrolls;

l, the input thickness (in mm);

v, the input speed (in m s^{-1}).

2. If, in the previous example, there is a blemish every 4.184 m, determine which pair of workrolls is at fault.

2.6 Coil feed line

A tinplate user installs a new coil feed line. The coil is placed on a mandrel of diameter 410 mm and the bottom of the mandrel is 650 mm from the floor. What is the maximum coil weight, in kg, that can be fed to the line if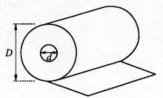

width of coil = 900 mm

density of steel = 7.85 gm cm^{-3}?

Maximum outside diameter is given by

$$650 + 650 + 410$$
$$= 1710 \text{ mm}$$
$$= 171 \text{ cm}$$

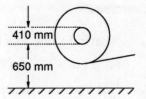

Now weight of coil

$$= \pi \frac{\left(D^2 - d^2\right)}{4} \times \text{width} \times 7.85 \text{ gm}$$

where

$D =$ outside diameter ($= 171$ cm)

$d =$ inside diameter ($= 41$ cm).

So weight of coil

$$= \frac{\pi}{4} \frac{(171 + 41)(171 - 41) \times 90 \times 7.85}{1000} \text{ kg}$$

$$= \frac{\pi}{4} \left(\frac{212 \times 130 \times 90 \times 7.85}{1000} \right) \text{ kg}$$

$$- 15293 \text{ kg}$$

Activity 2.15

What is the percentage increase in weight if the distance of the bottom of the mandrel from the floor is increased by 10 per cent?

Suppose that the thickness of the tinplate on the coil is 0.16 mm. What will be the length of the coil in metres?

If L denotes length, then equating the area of cross section gives

$$\frac{\pi}{4}(D^2 - d^2) = L \times \text{ thickness}$$

$$\Rightarrow \quad L \quad = \frac{\pi(D^2 - d^2)}{4 \times 0.00016}$$

$$= \frac{\pi(1.71^2 - 0.41^2)}{4 \times 0.00016}$$

$$\approx 13500 \text{ metres}$$

Exercise 2F

1.   What is the length for thickness

    (a) 0.1 mm   (b) 0.5 mm?

2.   Find a general solution for the length of coil in terms of D, d, and thickness ω.

2.7 Related data

In this section a variety of information is presented. The intention is to provide you with real data so that you can test hypotheses, look for correlations or simply analyze and present the trends in data. A list of questions you might investigate is given below; you will need to find further data for some of them.

The sources for the data given here are:

Annual Report and Accounts British Steel plc
British Steel plc
9 Albert Embankment
London SE1 7SN

World Steel in Figures
International Iron and Steel Institute (IISI)
Rue Colonel Bourg 120
B-1140
Brussels
Belgium

Questions

1. What are the trends in turnover for *British Steel* in the four given geographical areas? Illustrate these trends using appropriate diagrams. (Table 2.7.1)

2. From the data, estimate the profit for the next two years.
   Why are your estimates bound to be unreliable? (Table 2.7.1)

3. Is there strong correlation between the crude steel production for the 'World' and 'Western World'? (Table 2.7.2)

4. Is there any correlation between steel production and population of a country?
   (Tables 2.7.3(i) and (ii))

5. Describe the significant trends in changes of production for major steel-producing countries for 1980 to 1990. (Tables 2.7.3(i) and (ii))

6. Analyze the import / export data. What significant trends can you find?
   (Table 2.7.4)

7. Using the 1990 data, compare the efficiency of major steel-producing countries.
   (Table 2.7.5)

8. Analyze the trading data between different areas of the world. Does 'distance' play a significant part in trading? (Table 2.7.6)

2.7.1 Company data: *British Steel*

Group financial record

Profit and loss account	Mar 29 1986 £m	Mar 28 1987 £m	Apr 2 1988 £m	Apr 1 1989 £m	Mar 31 1990 £m	Mar 30 1991 £m	Mar 28 1992 £m
Mainstream turnover:							
United Kingdom	2 316	2 276	2 734	3 248	3 260	3 015	2 558
Rest of Europe	589	718	833	1 025	1 227	1 420	371
North America	195	211	237	295	292	262	278
Rest of World	179	256	312	338	334	344	391
Trading profit	91	190	424	656	708	323	17
Share of profits or related companies	22	21	23	35	76	42	6
Net interest and other income	(37)	(5)	8	42	94	81	22
Profit on ord. activities before taxation	42	177	419	593	733	254	55
Tax on profit on ordinary activities	(3)	1	(8)	(31)	(168)	60	20
Minority interests	(1)	–	(1)	(1)	(1)	(1)	(1)
Profit attributable to shareholders	38	178	410	561	564	193	34
Dividend	–	–	–	(100)	(165)	(175)	(90)
Profit retained	38	178	410	461	399	18	124
Earnings per share in pence	1.9*	8.9*	20.5*	28.0	28.2	9.65	1.7
Dividend per share in pence	–	–	–	5.0	8.25	8.75	4.5

* Based on net profit for year divided by the 2000 million ordinary shares issued in the Offer for Sale.

Balance sheet

	Mar 29 1986 £m	Mar 28 1987 £m	Apr 2 1988 £m	Apr 1 1989 £m	Mar 31 1990 £m	Mar 30 1991 £m	Mar 28 1992 £m
Fixed assets	2 226	2 414	2 506	2 630	3 088	3 447	3 449
Current assets	1 844	1 867	2 115	2 912	3 233	2 788	2 653
Creditors, provisions and minority interests	(1 216)	(1 247)	(1 187)	(1 632)	(2028)	(2 185)	(2 191)
Shareholders' funds	2 854	3 034	3 434	3 910	4 293	4 050	3 911

2.7.2 World steel production

World crude steel prodution 1970 –1990   (in million metric tons)

Year	World	Western world
1950	190	153
.	.	.
.	.	.
.	.	.
1960	336	241
.	.	.
.	.	.
.	.	.
.	.	.
1970	595	419
1971	.	396
1972	.	434
1973	697	491
1974	704	494
1975	643	424
1976	675	453
1977	676	443
1978	717	469
1979	747	497
1980	716	464
1981	707	459
1982	645	398
1983	664	407
1984	710	446
1985	719	451
1986	714	433
1987	737	449
1988	780	489
1989	785	497
1990	770	492

2.7.3(i) Major steel-producing countries (1989 and 1990)

Major steel-producing countries 1989 and 1990
million metric tons crude steel production

Country	1990 Rank	1990 tonnage	1989 Rank	1989 tonnage
USSR	1	154.4	1	160.1
Japan	2	110.3	2	107.9
United States	3	88.9	3	88.8
PR China	4	67.2	4	61.6
FR Germany	5	38.4	5	41.1
Italy	6	25.5	6	25.2
Republic of Korea	7	23.1	8	21.9
Brazil	8	20.6	7	25.1
France	9	19.0	10	18.7
United Kingdom	10	17.8	9	18.7
India	11	15.0	14	14.6
Czechoslovakia	12	14.9	11	15.5
Poland	13	13.6	13	15.1
Spain	14	12.9	16	12.8
Canada	15	12.3	12	15.5
Belgium	16	11.4	17	10.9
Taiwan (ROC)	17	9.7	19	9.0
Romania	18	9.7	15	14.4
Turkey	19	9.3	22	7.8
Mexico	20	8.7	20	7.9
South Africa	21	8.6	18	9.3
DPR Korea (E)	22	7.0	23	6.9
Australia	23	6.7	24	6.7
German Dem. Rep.	24	5.6	21	7.8
Netherlands	25	5.4	25	5.7
Sweden	26	4.5	27	4.7
Austria	27	4.3	26	4.7
Argentina	28	3.6	29	3.9
Yugoslavia	29	3.6	28	4.4
Luxembourg	30	3.6	30	3.7
Venezuela	31	3.2	32	3.2
Hungary	32	2.9	31	3.6
Finland	33	2.9	33	2.9
Indonesia (E)	34	2.6	35	2.4
Bulgaria	35	2.4	34	2.9
Egypt	36	2.1	36	2.1
Other Countries		18.4		17.9
World total		770.1		785.5

This table lists all countries producing more than two million metric tons of crude steel in either year shown.

2.7.3(ii) Major steel-producing countries (1980 and 1981)

Major steel-producing countries 1980 and 1981
million metric tons crude steel production

Country	1981 Rank	tonnage	1980 Rank	tonnage
USSR	1	149.0	1	147.9
USA	2	108.8	3	101.5
Japan	3	101.7	2	111.4
Federal Republic of Germany	4	41.6	4	43.8
China	5	35.6	5	37.1
Italy	6	24.8	6	26.5
France	7	21.3	7	23.2
Poland	8	15.6	8	19.5
United Kingdom	9	15.6	15	11.3
Czechoslovakia	10	15.2	11	14.8
Canada	11	14.8	9	15.9
Romania	12	13.5	12	13.2
Brazil	13	13.2	10	15.3
Spain	14	12.9	13	12.6
Belgium	15	12.3	14	12.3
India	16	10.8	16	9.5
Republic of Korea	17	10.8	18	8.6
South Africa	18	8.9	17	9.1
Australia	19	7.6	19	7.6
Mexico	20	7.6	21	7.1
German Democratic Republic	21	7.5	20	7.3
Democratic Republic of Korea	22	5.5	22	5.8
Netherlands	23	5.5	23	5.3
Austria	24	4.7	24	4.6
Yugoslavia	25	4.0	29	3.6
Luxembourg	26	3.8	25	4.6
Sweden	27	3.8	26	4.2
Hungary	28	3.6	28	3.8
Taiwan	29	3.1	27	4.2
Bulgaria	30	2.6	31	2.6
Argentina	31	2.6	30	2.7
Turkey	32	2.4	32	2.5
Finland	33	2.4	33	2.5
Venezuela	34	2.0	34	1.8
Other countries		12.5		13.4
World total		707.6		717.1

2.7.4 Major steel exporters and importers

1989 (million metric tons)

Rank	Total exports		Rank	Total imports	
1	Japan	20.2	1	United States	15.9
2	FR Germany	19.9	2	FR Germany	15.1
3	Belgium-Luxembourg	14.2	3	Italy	10.2
4	France	11.5	4	USSR (E)	10.2
5	Brazil	10.8	5	France	9.9
6	Italy	7.4	6	China	8.2
7	Republic of Korea	7.3	7	Japan	7.3
8	United Kingdom	6.7	8	Taiwan (ROC)	5.9
9	Netherlands	5.8	9	United Kingdom	5.5
10	USSR	5.3	10	German DR (E)	5.1
11	German DR (E)	4.5	11	Netherlands	4.9
12	Spain	4.4	12	Belgium-Luxembourg	4.7
13	United States	4.3	13	Republic of Korea	3.8
14	Canada	3.9	14	Iran	3.6
15	Turkey	3.6	15	Spain	3.2
16	Czechoslovakia	3.6	16	Turkey	2.9
17	Romania	3.3	17	Thailand	2.8
18	Austria	3.0	18	Hong Kong	2.8
19	Sweden	2.8	19	Canada	2.6
20	Poland	2.4	20	Switzerland	2.5

1980 (million metric tons)

Rank	Exports		Rank	Imports	
1	Japan	29.7	1	United States	13.7
2	FR Germany	19.0	2	FR Germany	11.4
3	Belgium-Luxembourg	13.8	3	France	8.8
4	France	11.3	4	USSR	8.0
5	Italy	6.8	5	Italy	7.0
6	USSR	6.5	6	China	5.6
7	Spain	4.7	7	United Kingdom	4.7
8	Republic of Korea	4.6	8	German DR	4.0
9	Netherlands	4.6	9	Netherlands	3.9
10	United States	3.8	10	Republic of Korea	3.2
11	Canada	3.5	11	Belgium-Luxembourg	3.1
12	Czechoslovakia	3.5	12	Mexico	2.5
13	United Kingdom	2.8	13	Taiwan	2.5
14	Austria	2.4	14	Saudi Arabia	2.2
15	Switzerland	2.1	15	Sweden	2.0
16	South Africa	2.0	16	Yugoslavia	2.0
17	Romania	2.0	17	Iran	2.0
18	Poland	1.9	18	Switzerland	1.9
19	German DR	1.9	19	Poland	1.7
20	Australia	1.6	20	Spain	1.7

2.7.5 Employment in the steel industry

Employment in the steel industry 1974 and 1984 to 1990 (thousand at end of year)

	1974	*1984*	*1985*	*1986*	*1987*	*1988*	*1989*	*1990*
Belgium	64	37	35	31	29	28	28	27
Denmark	2	2	2	2	2	2	2	2
France	158	85	76	68	58	53	40	46
FR Germany	232	153	151	143	133	131	130	127
Italy	96	76	67	66	63	59	58	56
Luxembourg	23	13	13	12	11	11	10	9
Netherlands	25	19	19	19	19	18	18	17
Portugal	4	6	6	6	6	5	4	4
Spain	89	69	54	51	45	41	38	36
United Kingdom	194	62	61	57	55	55	54	54
Total of above	887	522	484	455	421	403	392	378
Austria	44	35	34	32	29	26	22	21
Finland	10	13	10	10	10	10	10	10
Sweden	51	38	31	29	28	28	28	27
Yugoslavia	42	54	56	60	62	64	67	69
Canada	77	49	69	67	67	66	65	65
United States	521	268	238	220	203	212	206	204
Brazil	118	144	145	156	152	158	180	140
Australia	42	31	30	30	29	32	31	30
Japan	459	356	349	340	314	306	307	305
India	197	289	290	290	293	279	277	272
South Africa	100	111	110	111	112	111	111	112
Grand total	2548	1910	1846	1800	1720	1695	1686	1633

2.7.6 World steel

World steel trade by area 1989 (million metric tons)

Destination	EC	Other Western Europe	North America	Latin America	Africa	Middle East	Japan	Other Asia	Oceania	PR China	USSR & Eastern Europe	Total imports
						Exporting region						
EC	44.5	6.7	0.4	1.5	0.6	–	0.3	0.2	–	0.1	4.0	58.3
Other Western Europe	7.1	1.8	0.1	0.7	–	–	0.1	0.1	–	–	1.5	11.4
North America	5.6	1.2	3.6	3.2	0.1	–	3.6	2.0	0.3	–	0.3	19.9
Latin America	0.9	0.1	0.9	1.2	0.1	–	0.5	0.1	–	–	0.1	3.9
Africa	2.8	0.1	0.2	0.7	2.3	–	0.4	0.5	–	–	–	7.0
Middle East	1.7	0.4	0.1	0.3	–	0.6	1.0	2.0	–	–	–	6.1
Japan	0.2	0.2	0.7	1.1	0.1	–	–	4.4	0.1	0.1	0.5	7.3
Other Asia	2.6	0.4	1.9	4.6	–	–	9.0	3.2	0.6	0.7	–	23.0
Oceania	0.2	–	–	0.2	–	–	0.4	0.2	0.3	–	–	1.4
PR China and DPR Korea	0.9	0.3	0.3	1.0	–	–	3.9	0.2	–	–	2.2	8.8
USSR and Eastern Europe	4.5	2.4	0.2	0.1	–	–	0.9	–	–	–	12.7	20.8
Total exports	71.0	13.6	8.4	14.6	3.2	0.6	20.1	12.9	1.4	0.9	21.3	167.9

CHAPTER THREE

The Post Office

3.1 Postcodes

In 1837 *Rowland Hill* (who later became Chief Secretary to the Post Office), published a pamphlet called *'Post Office Reform: Its importance and practicality'* in which the main recommendations were that:

1. the postal service should be accessible to all, rich and poor; all citizens should be able to afford to use it;

2. the postal rate should be the same all over Britain. This was an important point, because it established the principle that the distance a letter was transported within these islands should not determine the postage rate;

3. the postage rate should depend on the weight of the letter (minimum rate one penny);

4. postage should be charged in advance. (This recommendation was not included in the Penny Postage Act of 1840.)

These recommendations led to the first postage stamps (Penny Black) being sold on 1 May 1840. Since that date there have been many changes to the system, but Rowland Hill's original recommendations are still in force.

What would be the effect if the second of his recommendations was dropped?

Much of the mail is now automatically sorted. Of particular importance has been the introduction of *postcodes* which were started in 1966 and now cover the whole of the UK. This alpha-numeric system, made up of between five and seven numbers and letters, was used firstly in order to divide the whole country efficiently and effectively into

areas   districts   sectors   units

and secondly, because people can remember a mixture of numbers and letters more easily than, for example, just numbers.

Do you think that this is true? How could you test this hypothesis?

Here is a typical postcode:

EX1 3PF

The first part, EX1, is called the *outward* code;
the second part, 3PF, the *inward* code.
Your first investigation will be to find how
many postcodes could exist with these rules.

Area	District	Sector	Unit
E X	1	3	P F
1 or 2 letters	any number 1 to 99	any digit 1 to 9	2 letters

Activity 3.1

(a) What is the maximum number of *areas* possible using postcodes in this form?

(b) For each area, how many *districts* could there be?

(c) For each district, how many *sectors* are possible?

(d) For each sector, how many *units* are possible?

Using your answers to these questions, how many *units* could there be in the whole country?

There are about 24 million household and business addresses in the UK.

Could each one have a unique postcode of the form above?

In fact, as is shown opposite, the Post Office uses

120   areas

There are in total about

2900   districts

9000   sectors

2 000 000   units

Activity 3.2

(a) On average, how many districts per area are used?

(b) How many sectors per district are used?

(c) How many units per sector are used?

(d) How many households/businesses are there per unit?

Why do you think the Post Office does not identify each address with a unique postcode?

Postcodes for the UK

AB	Aberdeen	GU	Guildford	PO	Portsmouth
AL	St Albans, Herts	HA	Harrow	PR	Preston
B	Birmingham	HD	Huddersfield	RG	Reading
BA	Bath	HG	Harrogate	RH	Redhill
BB	Blackburn	HP	Hemel Hempstead	RM	Romford
BD	Bradford	HR	Hereford	S	Sheffield
BH	Bournemouth	HU	Hull	SA	Swansea
BL	Bolton	HX	Halifax	SE	London SE
BN	Brighton	IG	Ilford	SG	Stevenage
BR	Bromley	IP	Ipswich	SK	Stockport
BS	Bristol	IV	Inverness	SL	Slough
BT	Belfast	KA	Kilmarnock	SM	Sutton
CA	Carlisle	KT	Kingston-upon-Thames	SN	Swindon
CB	Cambridge	KW	Kirkwall	SO	Southampton
CF	Cardiff	KY	Kircaldy	SP	Salisbury
CH	Chester	L	Liverpool	SR	Sunderland
CM	Chelmsford	LA	Lancaster	SS	Southend-on-Sea
CO	Colchester	LD	Llandrindod Wells	ST	Stoke-on-Trent
CR	Croydon	LE	Leicester	SW	London SW
CT	Canterbury	LL	Llandudno	SY	Shrewsbury
CV	Coventry	LN	Lincoln	TA	Taunton
CW	Crewe	LS	Leeds	TD	Galashiels
DA	Dartford	LU	Luton	TF	Telford
DD	Dundee	M	Manchester	TN	Tonbridge
DE	Derby	ME	Medway	TQ	Torquay
DG	Dumfries	MK	Milton Keynes	TR	Truro
DH	Durham	ML	Motherwell	TS	Cleveland
DL	Darlington	N	London N	TW	Twickenham
DN	Doncaster	NE	Newcastle-upon-Tyne	UB	Southall
DT	Dorchester	NG	Nottingham	W	London W
DY	Dudley	NN	Northampton	WA	Warrington
E	London E	NP	Newport, Gwent	WC	London WC
EC	London EC	NR	Norwich	WD	Watford
EH	Edinburgh	NW	London NW	WF	Wakefield
EM	Enfield, Middx	OL	Oldham	WN	Wigan
EX	Exeter	OX	Oxford	WR	Worcester
FK	Falkirk	PA	Paisley	WS	Walsall
FY	Blackpool	PE	Peterborough	WV	Wolverhampton
G	Glasgow	PH	Perth	YO	York
GL	Gloucester	PL	Plymouth	ZE	Lerwick

3.2 Coding and decoding postcodes

Letters which have postcodes included in the address are delivered with two rows of *phosphor dots* on the envelope. The upper row represents the inward code (e.g. 3PF) and the lower row the outward code (e.g. EX1). These dots are put on the letter automatically as soon as the letter arrives at a sorting office. The machine used is capable of reading most people's handwriting!

Sorting is then done automatically by an optical mark reader which reads from right to left. Each row has a *start* dot on the right which tells the optical reader to begin its scan.

The total number of dots in each row must be even, so an extra dot, a *parity* dot, is added at the far left when required. The letters are first sorted according to the outward portion of the code (e.g. EX1).

For example, if the letter shown opposite is posted in Bristol, it is sorted into the 'EX1' bag. This bag is then sent to the Exeter sorting office. This office then uses the inward code (3PF) for further automatic sorting, either into the sector, or individual units.

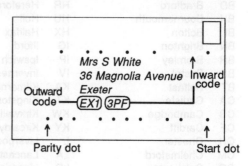

The coding of the inward and outward codes is based on the binary representation of numbers. This expresses any number in terms of powers of 2:

$$2^0 = 1, \quad 2^1 = 2, \quad 2^2 = 4, \quad 2^3 = 8, \quad \ldots$$

Example

$$53 = 32 + 16 + 4 + 1$$
$$= 1 \times 2^5 + 1 \times 2^4 + 0 \times 2^3 + 1 \times 2^2 + 0 \times 2^1 + 1 \times 2^0$$
$$= 110101_2$$

and

$$101110_2 = 1 \times 2^5 + 0 \times 2^4 + 1 \times 2^3 + 1 \times 2^2 + 1 \times 2^1 + 0 \times 2^0$$
$$= 32 + 8 + 4 + 2$$
$$= 46$$

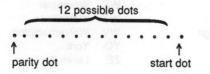

12 possible dots

parity dot   start dot

In fact, there are a possible 14 dots in each code (inward and outward) of which 12 are available for the outward code in the lower row, and 12 for the inward code in the top row. Remember that there is a start dot and a parity dot to make the total number even.

The code used is binary: $1 \equiv$ dot present, $0 \equiv$ no dot.

Activity 3.3

What is the largest number that could be coded using all 12 dots?

For the *outward code* there are 2900 different combinations used (see Section 3.1), and these are given a specified code, which is held in its entirety by a database in the computer machine that first codes the letters.

For the *inward code* there is an algorithm used to generate a unique number, say S, from the code, which is then coded into binary on the letter.

Activity 3.4

Without any restrictions, how many possible 'units' are there (see Section 3.1)? Can the 12-digit binary code be used?

In fact, the possible number of units that need to be coded is restricted by *not* using the letters   C, I, K, M, O and V in the inward code.

Activity 3.5

Given these restrictions, how many possible units are there?

Is the 12-digit binary code now appropriate?

In fact, the code number S is formed by using a base 20 representation. If

   N denotes the 'number'

   L_1 denotes the '1st letter'

   L_2 denotes the '2nd letter',

then the flow chart opposite gives the appropriate value of S.

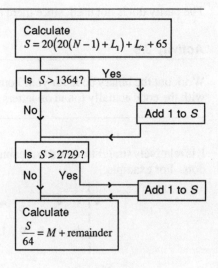

Calculate
$S = 20\big(20(N-1)+L_1\big)+L_2 +65$

Is $S > 1364$? — Yes

No

Add 1 to S

Is $S > 2729$?

No   Yes

Add 1 to S

Calculate
$\dfrac{S}{64} = M + \text{remainder}$

The table of weighting used is shown opposite for N, L_1 and L_2.

Table of weighting factors

A	0	P	10	O	10
B	1	Q	11	1	1
D	2	R	12	2	2
E	3	S	13	3	3
F	4	T	14	4	4
G	5	U	15	5	5
H	6	W	16	6	6
J	7	X	17	7	7
L	8	Y	18	8	8
N	9	Z	19	9	9

Example

Find the code number S for 3PF. Hence find its dot code.

Solution

Here $N = 3$, $L_1 = 10$, $L_2 = 4$, so

$$S = 20(20 \times 2 + 10) + 4 + 65$$

$$= 1069$$

Since $S < 1364$, you leave S as 1069 and

$$\frac{S}{64} = \frac{1069}{64} = 16 + \text{remainder } 45$$

Now

$$16 = 010000_2$$

$$45 = 101101_2$$

and the code is shown as

Parity dot	1 0 1 1 0 1	0 1 0 0 0 0	Start dot
	45	16	

The parity dot is not used, since there are (in total) six dots, which is already even.

Activity 3.6

Work out the binary code for your home or school/college. Check that it corresponds with the code actually found on letters received.

It is relatively straightforward to recapture the actual postcode, given the row of phosphor dots. For example,

1 1 0 0 1 0	0 1 0 1 1 0
50	22

Here $M = 0 \times 2^5 + 1 \times 2^4 + 0 \times 2^3 + 1 \times 2 + 1 \times 2^1 + 0 \times 2^0$

$\qquad = 22$

and remainder $= 1 \times 2^5 + 1 \times 2^4 + 0 \times 2^3 + 0 \times 2^2 + 1 \times 2^1 + 0 \times 20$

$\qquad\qquad = 50$

What value of S does this correspond to?

Now the coded value of S is given by

$\qquad S = 22 \times 64 + 50$

$\qquad = 1458$

but the original value will be $S = 1457$ (since it is > 1364 it will have been increased by 1 in the algorithm shown earlier in the flow diagram). But, in general,

$\qquad S = 20(20(N-1) + L_1) + L_2 + 65$

$\Rightarrow \quad S - 65 = N \times 20^2 + L_1 \times 20 + L_2 \times 1 - 400$

$\Rightarrow \quad S + 335 = N \times 20^2 + L_1 \times 20 + L_2 \times 1$

So the base 20 representation of $(S + 335)$ should give the values of N, L_1 and L_2.

For the example used, $1457 + 335 = 1792$ and

$\qquad 1792 = N \times 20^2 + L_1 \times 20 + L_2 \times 1$

$\Rightarrow \quad N = 4, \ L_1 = 9 \ \text{and} \ L_2 = 12$

which gives (see table of weighting factors) 4NR.

Activity 3.7

Find an envelope with a coded inward postcode and (without looking) determine the actual postcode. Check your answer.

What is the significance of the two S values 1364 and 2729 given in the algorithm?

Postcodes are crucial to the efficiency of the letter-posting/delivery system. Over 80 per cent of letters now carry the postcode but the postcodes omitted from the other 20 per cent of letters have to be looked up and marked on, to ensure that the letters are automatically sorted.

Exercise 3A

1.   Find the value of S and the dot codes for
      (i) EX4 9JN   (ii) SS14 1AP   (iii) CB1 4QQ

2.   Decode these final S values into postcodes.
      (i) 1000   (ii) 3720   (iii) 1365

3.   What is the largest value of S that can be obtained using the dot representation?
      What is the relationship between this number and 1364 and 2728?

3.3 Postal rates

The current UK postal rates are shown below. These took effect from 1 November 1993.

Weight not over	First Class	Second Class	Weight not over	First Class	Second Class
60 g	25p	19p	500 g	£1.25	98p
100 g	38p	29p	600 g	£1.55	£1.20
150 g	47p	36p	700 g	£1.90	£1.40
200 g	57p	43p	750 g	£2.05	£1.45
250 g	67p	52p	800 g	£2.15	
300 g	77p	61p	900 g	£2.35	Not admissible over 750 g
350 g	88p	70p	1000 g	£2.40	
400 g	£1.00	79p	Each extra 250 g or part thereof 65p		
450 g	£1.13	89p			

Royal Mail aims to deliver (Monday to Saturday) first class letters the day after collection and second class letters by the third working day after collection.

Why do you think the Royal Mail has a two-tier service?

Activity 3.8

Draw an accurate graph to illustrate both the first and second class postal rate data.
Is the percentage saving using second class rather than first class similar at all weights?

What sort of functions have you drawn in the activity above?

Stamps are currently available for the following amounts:

1p, 2p, 3p, 4p, 5p, 6p, 10p, 19p, 20p, 25p, 29p, 30p, 35p,

36p, 38p, 41p, 50p, £1.00, £1.50, £2.00, £5.00, £10.00

Activity 3.9

Check the minimum number of stamps needed to make each of the values in the table of postal rates. Can any of the stamps currently on sale be deleted without increasing the number of stamps needed to make up the stated values?

Can you suggest a better system of stamp values?

The Post Office has been continually replacing equipment with new automatic coding and sorting machines which have increased the efficiency of the delivery of letters. As stated earlier, the aim is to deliver all first class letters the day after posting (although in far distant offices, this is not always practical).

Activity 3.10

Test the efficiency of your local sorting office by posting first class, say 10 or 20 letters to local addresses. Note how many arrive the following day. Repeat the experiment sending letters to more distant places.

The Post Office publishes efficiency data related to the area of location of posting. (See Section 3.6.6.)

Is your data in line with the published performance results?

Activity 3.11

Use hypothesis testing to see if your data is in agreement with the published performance results.

3.4 Maximizing volume

You do not usually need to consider weight and size limitation when using the Royal Mail but, for some businesses, this is of vital importance. There is no *weight* limitation for first class letter post but second class has a weight limit of 750 g. There are, though, strict limitations on *size*.

The following is an extract from the *Post Office Guide* (£1 from main Offices) for inland letter post.

> 610mm (2 feet) in length, 460mm (1 foot 6 inches) in width and 460mm (1 foot 6 inches) in depth; or if made up in the form of a roll, 1.040m (3 feet 5 inches) for the length and twice the diameter combined, and 900mm (2 feet 11 inches) for the greatest dimension.

What is the maximum volume of a cuboid which can be posted?

The situation is not so straightforward for 'rolls', i.e. cylinders. Suppose a cylinder has diameter x m and length l m. The constraints on the roll so that it can be posted are

$$l + 2x \;\leq 1.04$$

$$l \quad\;\; \leq 0.9$$

$$x \quad\;\; \leq 0.9$$

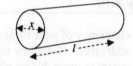

Taking an extreme case, namely $l = 0.9$, you can see that

$$l = 0.9 \Rightarrow 2x \leq 1.04 - 0.9$$

$$2x \leq 0.14$$

To maximize the volume with this value of l, you clearly take $x = 0.07$. This gives a volume of

$$V = \pi \left(\frac{d}{2}\right)^2 l$$

$$= \pi \left(\frac{0.07}{2}\right)^2 0.9$$

$$V \approx 0.00346 \text{ m}^3$$

Why is the extreme case $x = 0.9$ m not feasible?

Activity 3.12

Either show that the dimensions used above ($l = 0.9$ m, $x = 0.07$ m) give the roll of maximum volume that can be posted, or find dimensions that give a greater volume.

Suppose you assume that $l = 0.9$ m is *not* the optimum answer. Then it is clear that the inequality $l + 2x \leq 1.04$ will in fact be an equality $l + 2x = 1.04$.

The volume will be given by

$$V = \pi\left(\frac{x}{2}\right)^2 l = \pi\left(\frac{x}{2}\right)^2 (1.04 - 2x)$$

$$= \frac{\pi}{4}\left(1.04x^2 - 2x^3\right)$$

Using differentiation to find the stationary values of V gives

$$\frac{dV}{dx} = \frac{\pi}{4}\left(2.08x - 6x^2\right)$$

$$= \frac{\pi}{4}x(2.08 - 6x)$$

$$= 0 \text{ when } x = 0 \text{ or } x = \frac{2.08}{6} = \frac{26}{75} \ (\approx 0.347)$$

Will this value of x give a maximum value?

One way to check is to evaluate $\dfrac{d^2V}{dx^2}$ at this value of x.

$$\frac{d^2V}{dx^2} = \frac{\pi}{4}(2.08 - 12x)$$

$$= -0.52\pi \text{ at } x = \frac{26}{75}$$

$$< 0$$

Hence there is indeed a maximum value of V at $x = \dfrac{26}{75}$.

Its actual value can be readily calculated. First, find l from

$$l + 2x = 1.04 \implies l = 1.04 - 2 \times \frac{26}{75}$$

$$= \frac{26}{75}$$

What do you notice about the values of l and x?

The maximum volume is hence given by

$$V = \pi \left(\frac{13}{75}\right)^2 \frac{26}{75} \approx 0.0327 \text{ m}^3$$

Whilst this result is, in general, of academic rather than practical interest, another more realistic problem is that represented by seeing if it is possible to post a picture frame of size 610 mm × 750 mm, using letter post.

Assuming that the picture and wrapping are of negligible thickness you can use a cuboid box to package the frame – but is it acceptable for posting?

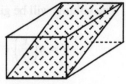

Activity 3.13

Find the dimensions of a cuboid required to package the frame, such that it meets the constraints for letter posting.

Activity 3.14

Any parcel that can be carried by post in the UK must obey the following constraints

maximum dimensions allowed
length 1.5 m
length and circumference combined 3 m

Find:
- (a) the maximum volume that can be carried if it is in the shape of a cuboid;

- (b) the maximum volume that can be carried if it is in the shape of a cylinder.

3.5 Optimum delivery and collection routes

The Royal Mail is continually faced with both delivery and collection route problems. The objectives, of course, are to find the appropriate routes which minimize costs.

The cost of delivery will depend on such factors as

(i) mode of transport   (ii) distance travelled   (iii) time taken.
     (e.g. walking/cycling/van)

The problems of *collection* are first discussed and analysed, since this is intrinsically an easier problem to cope with. (The Royal Mail in fact uses a sophisticated program called TRANDOS–Transport Disc Operating System, to find optimum solutions.)

The collection problem is essentially a 'travelling salesman' problem, in which you start and finish at the same point (i.e. the sorting office) and wish to visit a number of specified points (i.e. post boxes) at least once on your route. A possible criteria to take is to ask for the route of minimum distance.

Will such a route always also be the route of minimum time?

A very simple example is shown here.

The routes on a housing estate are represented by lines.

Each post box is indicated by an asterisk * and
a letter (A to E).

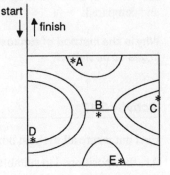

Activity 3.15

Given that the sketch map is to scale, what is the route with minimum distance passing each box and returning to the start point?

Even this simple problem is not easy to solve – the real problems faced by the Royal Mail even in a small town involve many more boxes and more than one collection van! Hence the need for a computerized solution, although we can understand the basis for the solution by looking at some further, more manageable problems.

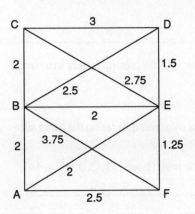

This diagram represents a simple network in which there are post boxes at A, B, C, D, E and F. (A is close to the sorting office.) The distances in km between each box along available roads are also shown.

Activity 3.16

In the last diagram, find the optimum route for a collection van starting at the sorting office at A.

Unfortunately, there is currently no neat method of solving this problem which guarantees obtaining the optimum solution on every occasion. The only method available is called the *method of exhaustion*, in which the total distances for every possible path are found and compared.

Why is the method of exhaustion not particularly useful when there are many post boxes to be visited?

Activity 3.17

Find out where all the post boxes are in your local area.

Are they positioned in suitable places?

Find the optimum route for a collection van to take.

We now turn to the problems of optimum delivery.

Why is delivery more complicated than collection?

For delivery in congested housing areas, you need to consider the question of whether the postal worker should walk up and down the street or whether it is more efficient to keep crossing the road.

Activity 3.18

Design a simple model to help see the most efficient way of delivering letters in a street with houses along each side.

Two possible strategies are:

(a) deliver along one side (houses 1, 3, ..., 13), cross over and deliver on the other side (houses 14, 12, ..., 2);

(b) deliver alternately to houses on each side of the road (houses 1, 2, 3, 4, ..., 14).

What factors will influence the choice of delivery strategy?

From now on, you can assume that the delivery strategy will require the postal worker to travel in just one direction along each road.

Activity 3.19

Assuming that the postal worker travels along each road at least once for the delivery, find the minimum total distance travelled for each of the networks representing housing estates shown below. In each case start and finish at the node marked ●. (Diagrams are not drawn to scale.)

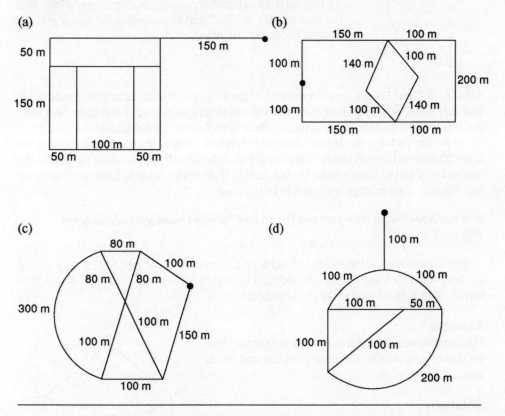

By now, you should be getting a feel for the general solution of such problems. (They are known by the title *Chinese postman problems*, due to the fact that these were first solved by Chinese mathematicians.)

In all your solutions to the problems in Activity 3.18 you will have noted that often one section of road has to be repeated in the optimum route.

What is the key factor determining if the road (or section of road) has to be repeated?

This problem is essentially the same as the *Königsburg bridge problem*, which was investigated by the people of that town on Sunday afternoons when they tried to find a route, starting and finishing at the same place, passing over each bridge once and only once.

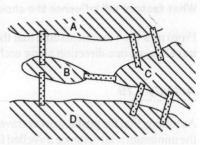

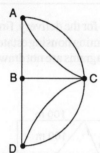

To find out if this is possible, redraw the map as a network, with the land masses, A, B, C and D becoming the nodes and the bridges becoming the edges.

All of A, B, C and D have an *odd* number of edges joining them and each must be used once, and only once. So, if you start at A, you will return on a second edge and leave on a third but you will not be able to finish there – there is no further edge available.

So the key factor is the *degree* of the node (vertex) – that is, the number of edges at the node. If a network has all nodes with even degrees, then it will be possible to transverse the network without repeating any edge (i.e. road). If the network has at least one node with odd degrees, then an edge will need to be repeated.

It is not possible to have just one (or an odd number) node with odd degrees. Why not?

Having ascertained the number of odd degree vertices, then the problem remains to find the best way of joining up the odd nodes so as to give a minimum total extra distance to travel. This is illustrated in the next example.

Example

Find an optimum route for the postal worker problem for the network shown. The route must start and finish at B.

Solution

The vertex degrees are

A	B	C	D	E	F
3	2	3	3	3	4

So there are four nodes with odd degrees. The problem is to find the best way of repeating two edges, effectively making each node even.

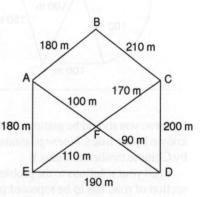

There is one obvious way, namely

 repeat A to E and C to D, giving $180 + 200 = 380$ m

extra distance to cover.

This gives the network shown.

There are now many ways to traverse the network, for example

 B - C - D - F - C - D - E - A - F - E - A - B.

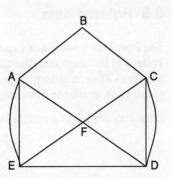

Activity 3.20

Find different ways to traverse this network with the same total length.

Exercise 3B

For each of the networks shown below, find the route of minimum total distance which passes along each edge at least once. Start and finish at the node marked with an arrow.

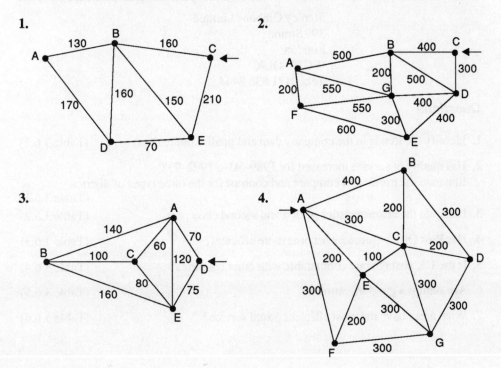

3.6 Related data

The Post Office publishes a variety of information concerning postal rates, postal targets, volume of business, conditions of carriage, etc. Much of this can be freely collected from any Post Office, and the precise rates of posting can be found in the publication *Post Office Guide* (£1), available from larger Post Offices.

Enquiries concerning educational publications should be addressed to

> Post Office Educational Services
> PO Box 145
> Sittingbourne
> Kent
> ME10 1NH

More general enquiries should be addressed to

> The Post Office Public Affairs Division
> 130 Old Street
> London
> EC1V 9 HQ
> Tel: 0171 490 2888

Details of the values of past stamps can be obtained from catalogues produced by

> Stanley Gibbons Limited
> 399 Strand
> London
> WC2R OLX
> Tel: 0171 836 8444

Questions

1. Identify the trends in the company data and predict future data.   (Table 3.6.1)

2. Has quality of service increased for 1989–90 to 1992–93?
   Illustrate the trends and compare and contrast for the three types of district.
   (Table 3.6.1)

3. Illustrate the characteristics of first and second class post.   (Table 3.6.2)

4. Has Post Office Counters become more efficient?   (Table 3.6.3)

5. Is the UK postage rate comparable with other countries?   (Table 3.6.4)

6. Are stamps a good investment?   (Table 3.6.5)

7. What areas have the most efficient postal service?   (Table 3.6.6)

3.6.1 Company data: The Post Office

Profit and loss account	*87–88* £m	*88–89* £m	*89–90* £m	*90–91* £m	*91–92* £m	*92–93* £m
Turnover	3,791	3,915	4,459	4,719	5,149	5,345
Profit before interest	171	116	78	106	231	230
Net interest receivable	26	47	38	47	16	53
Profit before exceptional items	197	163	116	153	247	283
Exceptional items	–	–	–	(106)	–	–
Profit before taxation	197	163	116	47	247	283
Taxation	(76)	(61)	(44)	(16)	(95)	(96)
Extraordinary items	–	–	(69)	–	–	–
Profit for the financial year	121	102	3	31	152	187
Dividend	(2)	(2)	–	–	–	–
Retained profit	119	100	3	31	152	187

Staff at each year-end	*87–88* '000	*88–89* '000	*89–90* '000	*90–91* '000	*91–92* '000	*92–93* '000
Royal Mail	162	167	172	171	166	158
Parcelforce	10	12	12	13	12	13
Post Office Counters	18	18	17	17	15	15
Subscription Services	1	1	1	1	1	1
Corporate	3	3	3	4	3	3
Girobank	6	6	6	–	–	–
	200	207	211	206	197	190

Quality of first class letter service: % of letters delivered next day

	89–90	*90–91*	*91–92*	*92–93*
Overall	78.1	85.5	89.8	91.9
Within district	89.3	92.8	94.3	95.8
Adjacent districts	80.6	87.2	91.3	92.9
Distant districts	68.0	78.8	85.4	86.6

Quality of second class letter service: % of letters delivered within three days

	89–90	*90–91*	*91–92*	*92–93*
Overall	94.0	96.4	97.8	98.4
Within district	98.2	98.8	99.3	99.5
Adjacent districts	94.0	96.9	98.4	98.8
Distant districts	89.8	93.6	95.0	96.4

3.6.2 Royal Mail data

		87–88	*88–89*	*89–90*	*90–91*	*91–92*
Inland						
No. of first class letters	m	6,055	6,067	6,546	6,659	6,418
No. of second class letters	m	6,936	7,137	8,173	8,647	8,961
Total[1]	m	12,991	13,204	14,719	15,306	15,379
Growth index[1]		100.0	103.0	112.9	115.8	116.7
Inflation-adjusted tariff index		100.0	97.9	96.2	95.3	99.4
Letters postcoded[2]	%	72.4	77.0	78.5	80.3	82.7
International						
No. of letters posted	m	577	538	575	596	659
Growth index		100.0	93.6	99.1	102.5	113.9
Revenue index (at constant prices)		100.0	98.5	102.4	105.6	103.5
Air letter quality of service						
Inward[3]	%	81.1	81.4	79.0	89.4	92.7
Outward[4]	%	88.6	85.5	83.4	92.6	96.1
Operational						
Operating hours	m	352	356	372	374	360
Overtime hours	m	46	45	46	41	33
Days lost through industrial action	'000	66.8	1,164.5	42.0	18.5	1.2
Road transport						
No.of vehicles	'000	29.2	29.0	31.4	29.2	28.9
Total mileage	m	373	353	410	443	420

Notes:

1 The growth index excludes the effect of the change in traffic derivation from 1990–91, the effect of different lengths of accounting year, and for inland, the General and European Elections.

2 Postcode usage is sampled continuously. The figures show usage by the year-end. Response Services items are not included in the sample.

3. The inward performance figures show the percentage of air letters received in time for general night mail despatches and delivered on the day after arrival in the UK. The target for 1991–92 was 90% (89% in previous years).

4 The outward figures represent the percentage of air letters ready for despatch by mid-day on the day after posting in the UK. The target for 1991–92 was 93% (90% in previous years).

3.6.3 Post Office Counters

	87–88	88–89	89–90	90–91	91–92
Number of Post Offices					
Directly operated offices	1,499	1,493	1,339	1,167	1,019
Agency offices	19,572	19,537	19,532	19,471	19,141
Value of transactions (£m)					
Girobank services					
Deposits	32,223	34,262	36,590	40,648	45,056
Withdrawals	2,333	2,311	2,367	2,532	2,750
DSS orders	5,976	5,280	4,661	5,218	7,025
Other services	2,265	2,862	2,805	2,943	3,090
	42,797	44,715	46,423	51,341	57,921
Pensions and allowances					
Child benefit payments	4,738	4,694	4,437	4,539	4,978
National insurance and pensions	24,954	26,461	28,448	31,765	35,892
	29,692	31,155	32,885	36,304	40,870
Licences issued					
Motor vehicles	2,277	2,403	2,537	2,668	2,672
Television	862	934	1,031	1,024	1,058
	3,139	3,337	3,568	3,692	3,730
National Savings services					
NSB deposits	2,152	1,988	1,841	1,874	1,566
NSB withdrawals	753	763	802	813	773
Other services	421	494	303	346	307
	3,326	3,245	2,946	3,033	2,646
Telephone bills paid	2,149	2,068	2,132	2,283	2,350
Postage stamps	818	835	905	986	1,069
Savings and other stamps	587	619	664	716	771
Postal orders	278	275	288	284	299
Other services	255	249	261	282	295
	83,041	86,498	90,072	98,921	109,951

3.6.4 Worldwide postal rates

Minimum letter postage for a selection of EC countries (internal)

Country	Postage
Belgium	15 BF
Denmark	3.75 DK
France	2.8 francs
Germany	1 DM
Greece	60 dr
Ireland	32p (Irish pence)
Italy	750 lire
Luxembourg	14 francs
Netherlands	0.8 florin
Portugal	75 escudos
Spain	28 pesetas
UK	19p (second class)
	25p (first class)

3.6.5 Value of stamps

| | *1994* | *1984* | *1974* | *1964* | *1954* |
	£	£	£	£	£
Great Britain 1840 1d black used (SG2) 'Penny Black'	150.00	125.00	12.00	3.75	3.00
Great Britain 1929 £1 Postal Union Congress (SG438) used	400.00	475.00	65.00	11.00	9.00
Australia 1932 5/- Sydney Harbour Bridge unused (SG143)	375.00	450.00	40.00	8.00	8.00
Canada 1851 12d black (SG4) used	40,000.00	38,000.00	9,000.00	2,750.00	800.00
Falkland Islands 1933 Centenary set of 12 unused	2,250.00	2,250.00	270.00	120.00	65.00
India 1854 4a (SG17) unused	3,750.00	2,000.00	160.00	85.00	85.00
Straits Settlements 1910 $500 (SG169) unused	50,000.00	16,000.00	10,000.00	5,250.00	750.00
New Zealand 1931 Health set of 2 (SG546/7) unused	150.00	180.00	26.00	9.50	14.00
Rhodesia 1910/13 £1 error of colour (SG1666) unused	8,500.00	8,500.00	1,500.00	120.00	60.00
Cape of Good Hope 1861 4d vermilion 'Woodblock' (SG14e) used	40,000.00	25,000.00	3,500.00	1,850.00	1,000.00

3.6.6 First class letters: performance results for 1993/94

The table below shows the percentage of first class letters delivered the next day, for local (*L*), neighbouring (*N*) and distant (*D*) postings in the specified postcode areas. Survey period July 1993 – June 1994.

Anglia		L %	N %	D %	London		L %	N %	D %
AL	St Albans	97	92	90	NW	North West London	91	89	83
CB	Cambridge	97	95	89	SE	South East London	95	92	87
CM	Chelmsford	97	95	88	SW	South West London	92	89	83
CO	Colchester	96	95	89	WI	West London WI	94	92	89
EN	Enfield	95	91	88	W2-14	West London W2-14	93	91	85
IG	Ilford	96	92	87	WC	West Central London	94	93	89
IP	Ipswich	95	94	89	**South Central**				
LU	Luton	95	94	90	GU	Guildford	96	91	85
MK	Milton Keynes	96	91	87	HA	Harrow	94	90	86
	Norwich	97	95	89	HP	Hemel Hempstead	97	92	87
PE	Peterborough	96	93	88	OX	Oxford	97	95	90
RM	Romford	92	92	82	PO	Portsmouth	95	94	86
SG	Stevenage	94	95	89	RG	Reading	95	92	87
SS	Southend-on-Sea	94	93	87	SL	Slough	98	93	88
WD	Watford	97	95	88	SN	Swindon	95	88	88
North East					SO	Southampton	96	93	90
BD	Bradford	97	94	89	SP	Salisbury	97	93	89
DH	Durham	95	94	87	UB	Uxbridge	96	92	85
DL	Darlington	96	95	88	**South East**				
DN	Doncaster	96	96	88	BN	Brighton	97	94	88
HD	Huddersfield	96	95	88	BR	Bromley	96	92	89
HG	Harrogate	95	91	82	CR	Croydon	96	90	91
HU	Hull	97	97	89	CT	Canterbury	97	95	88
HX	Halifax	94	93	86	DA	Dartford	95	92	87
LN	Lincoln	97	94	89	KT	Kgstn upon Thames	91	88	83
LS	Leeds	96	94	88	ME	Medway & Mdstn	94	88	81
NE	Newcastle	96	90	87	RH	Redhill	94	94	87
S	Sheffield	94	93	87	SM	Sutton	94	88	80
SR	Sunderland	96	93	88	TN	Tonbridge	94	91	84
TS	Cleveland	94	92	88	TW	Twickenham	98	95	91
WF	Wakefield	97	93	86	**North Wales and North West**				
YO	York	97	93	87	BB	Blackburn & Burnley	95	92	81
London					BL	Bolton	95	95	87
E	East London	93	90	84	CA	Carlisle	95	91	87
EC	City of London	97	95	90	CH	Chester & Deeside	96	94	86
N	North London	95	90	83	CW	Crewe	96	95	88

3.6.6(continued) first class letters: performance results for 1993/94

North Wales and North West		L %	N %	D %
FY	Fylde	96	94	89
L	Liverpool & Wirral	93	89	85
LA	Lancaster	96	90	87
LD	Llandrindod Wells	93	87	78
LL	Llandudno	97	89	88
M	Manchester	92	91	86
OL	Oldham	94	93	87
PR	Preston	93	93	82
SK	Stockport	96	92	84
SY	Shrwsbry & M.Wales	96	89	86
TF	Telford	96	93	83
WA	Warrington	97	94	86
WN	Wigan	96	94	91
Midlands				
B	Birmingham	96	92	89
CV	Cvntry & Wrwkshr	97	94	89
DE	Derby	98	94	92
DY	Dudley	93	90	82
LE	Leicester	94	89	86
NG	Nottingham	95	92	88
NN	Northamptonshire	96	91	87
ST	Stoke-on-Trent	95	93	86
WS	Walsall	95	92	84
WV	Wolverhampton	97	93	82
South Wales and South West				
BA	Bath	96	95	88
BH	Bournemouth	98	94	88
BS	Bristol	95	89	86
CF	Cardiff	95	87	79
DT	Dorchester	98	95	92
EX	Exeter	96	91	87
GL	Gloucester	96	89	85
HR	Hereford	96	95	90
NP	Newport	92	90	76
PL	Plymouth	94	90	85
SA	Swansea	97	93	88
TA	Taunton	96	92	86
TQ	Torquay	96	96	89
TR	Truro	96	91	87
WR	Worcester	98	92	91

Scotland and Northern Ireland		L %	N %	D %
AB	Aberdeen	96	93	86
BT	Northern Ireland	97	89	86
DD	Dundee	96	92	80
DG	Dumfries	96	90	86
EH	Edinburgh	96	94	88
FK	Falkirk	96	93	81
G	Glasgow	95	90	85
IV	Inverness	94	87	75
KA	Kilmarnock	97	95	82
KW	Kirkwall	90	90	73
KY	Kirkcaldy	96	96	84
ML	Motherwell	97	96	86
PA	Paisley	94	94	83
PH	Perth	95	93	82
TD	Borders	95	92	86
ZE	Lerwick	95	89	82

CHAPTER FOUR

Esso

4.1 Oil pipelines

Esso uses its Fawley Refinery (near Southampton) to make most of the oil products it sells in the UK. Over 85 per cent of these products are distributed by a network of underground pipelines to a number of depots with storage facilities throughout the country. From there, the fuel is transported to customers, such as petrol stations. The pipeline is expensive to build but once laid, it is essentially invisible, environmentally friendly, safe and inexpensive to maintain. The current pipeline network and depots are shown on the map below.

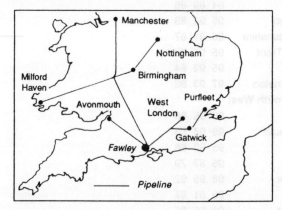

The flow in the pipelines is continuous for 24 hours every day, and different products follow one another down the pipeline. It has been found that these different products will not mix, so long as the speed is at least 6 km/hour. Consequently only a small interface between different products is maintained. Speed is controlled by having pumping stations at regular intervals along the line.

One of the many mathematical problems associated with pipelines occurs at the junctions. The flow *must* be continuous and the speeds need to be adjusted according to the *diameter* of the pipes.

As a simple example, consider the junction shown below.

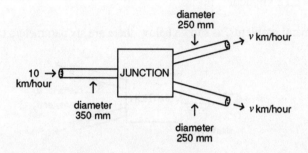

If the inflow speed is 10 km/hour and the pipe diameter is 350 mm, the volume of oil arriving at the junction in one hour is given (in m³) by

$$v = 10000 \times \text{ area of cross-section}$$

$$= 10000 \times \pi \left(\frac{0.350}{2}\right)^2$$

$$= 306.25\,\pi\, \text{m}^3$$

If the outflow speeds are equal and the diameters each 250 mm, we can find the outflow speed, since no oil must be lost or gained. Let v be the outflow speed in km/hour in each pipe.

What is the outflow in one pipe in one hour?

Care must be taken with units. The outflow in one pipe is given by

$$\text{speed} \times \text{cross-section} \quad = (1000\,v) \times \left(\pi \left(\frac{0.250}{2}\right)^2\right)$$

$$= 15.625\ \pi v$$

Hence

$$306.25\,\pi = 15.625\,\pi v + 15.625\,\pi v$$

This gives

$$v = 9.8 \ \text{km/hr}$$

In the same way, if you are given an outflow speed of, say, 6 km/hour, this is sufficient information to determine the inflow speed, say v km/hour. You must have

$$1000\,v \times \pi \left(\frac{0.350}{2}\right)^2 = 2 \times 6000 \times \pi \left(\frac{0.250}{2}\right)^2$$

This gives

$$v = 6.12 \text{ km/hour}$$

In the most general problems, as shown below, there are six parameters to be chosen.

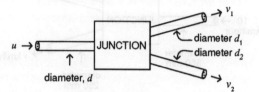

How many equations are there to be satisfied?

The general formula gives

$$d^2 u = d_1^2 v_1 + d_2^2 v_2$$

Activity 4.1

(a)   Check that the two problems above satisfy this equation.

(b)   Investigate the sensitivity of the solutions, from the equation, to changes in the parameters.

Activity 4.2

Extend the analysis above to the configuration illustrated below.

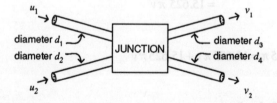

Are the following situations feasible, and do they have unique solutions?

(i) $u_1 = u_2 = 10 \text{ km/hour}$, $d_1 = d_2 = 250 \text{ mm}$, $d_3 = d_4 = 500 \text{ mm}$, $v_2 = 0$

(ii) $u_1 = 20 \text{ km/hour}$, $u_2 = 10 \text{ km/hour}$, $d_1 = d_2 = 250 \text{ mm}$, $v_1 = v_2$

(iii) $u_1 = u_2 = 5 \text{ km/hour}$, $d_1 = d_2 = 500 \text{ mm}$, $d_3 = d_4 = 250 \text{ mm}$, $v_1 = 15 \text{ km/hour}$

4.2 Value for money

The UK has been fortunate to be able to move from being an *importer* of oil to a net *exporter*. This has been due to the exploration and extraction of oil from the North Sea.

But has oil independence affected the price at the petrol pumps?
Has petrol become relatively cheaper over the past decade?

Year	Retail Price Index	Pump price (pence per litre)	Tax payable (pence per litre)
1972	23.0	7.70	4.95
1973	25.1	7.70	4.95
1974	29.1	9.24	4.95
1975	36.1	15.95	4.95
1976	42.1	16.83	6.60
1977	48.8	17.49	7.70
1978	52.8	16.76	6.60
1979	59.9	17.50	8.10
1980	70.7	26.39	10.00
1981	79.1	29.05	13.57
1982	85.9	35.02	15.54
1983	89.8	36.70	16.30
1984	94.3	40.35	17.16
1985	100.0	41.54	17.94
1986	103.4	41.63	19.38
1987	107.7	38.42	19.38
1988	113.0	36.79	19.38
1989	121.8	32.14	19.38
1990	133.3	40.92	22.48
1991	141.1	45.13	25.85

The *Retail Price Index* (RPI) is measured relative to a value of 100 at the beginning of 1985. To compare the values, adjust all the 4-star prices so that the 1972 value is 23.0.

This means multiplying each price by $\dfrac{23}{7.7}$.

So the 1973 index becomes

$$7.70 \times \left(\frac{23}{7.7} \right) = 23$$

and the 1974 index is

$$9.24 \times \left(\frac{23}{7.7} \right) = 27.6$$

Activity 4.3

Continue these calculations to see if a trend occurs.

To find out if petrol is good value for money, you can draw a graph of the adjusted 4-star price index against RPI.

Activity 4.4

Draw a graph of *4-star price index* against the RPI for 1972–1991. If the price had kept up with the inflation rate, then you would expect a line through the origin of slope 45°. From your graph, what do you conclude?

This is not a fair measure for the oil companies, since the taxes paid to the government have also been increasing.

Activity 4.5

Repeat the process, but this time use the *4-star price* minus *tax payable*. What are your conclusions now?

4.3 Dipstick problem

Petrol stations very rarely run out of fuel. This is due partly to regular deliveries but also, more importantly, to efficient stock control. Each type of fuel (4-star, unleaded, diesel) is stored in an underground tank and the amount in each tank is carefully monitored using some form of dipstick.

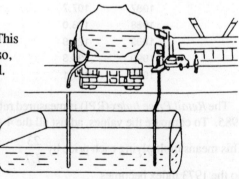

It is easy to measure the *height*, say *h*, of fuel in a tank. However, the volume will be proportional to the *cross-sectional area*, not the height.

Suppose the cross-section is a circle (it is, in fact, usually elliptical, but a circle is an approximation). You need to find the relationship between area, *A*, and height, *h*, and so provide a ready reckoner to convert height to area.

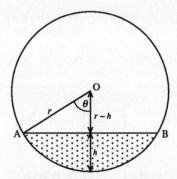

For simplicity, take $r = 1$ m. For values of h from 0 to 1, in steps, you need to find the angle θ and the area of fuel. From the diagram,

$$\cos\theta = \frac{r-h}{r} = \frac{1-h}{1} = 1-h$$

The cross-section of area of fuel is found by first finding the cross-section OAB of the circle. This is given by

$$\pi \times \frac{2\theta}{360} = \frac{\pi\theta}{180}$$

What is the area of triangle OAB?

To find its area, note that the length $AB = 2\sin\theta$

So

$$\text{area of triangle OAB} = \frac{1}{2} \times (2\sin\theta) \times (1-h)$$
$$= (1-h)\sin\theta$$

Hence

$$\text{area of fuel} = \frac{\pi\theta}{180} - (1-h)\sin\theta$$

As a fraction of the total cross-section of the tank, that is π,

$$\text{area fraction } A^* = \frac{\theta}{180} - \frac{(1-h)}{\pi}\sin\theta$$

Activity 4.6

Using this function, complete the table below.

h	$\theta°$	Area fraction
0	0	0
0.1	25.84	0.019
...	...	...
...	...	...
...	...	...
1.0	90.00	0.500

Use this information to draw a graph of A^* (vertical axis) against h (horizontal axis), or alternatively use a graphics calculator directly.

You should now have sufficient information to complete the lower half of a dipstick by calibrating the area fraction.

Activity 4.7

Complete the lower half of a dipstick using the data from Activity 4.6.

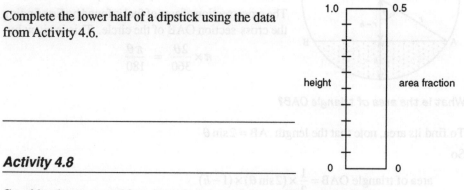

Activity 4.8

Consider the same problem, but when the tank is an ellipse of the form

$$\frac{x^2}{a^2} + \frac{y^2}{b^2} = 1$$

4.4 Depot location

In Section 5.1 in the next chapter, we consider the problem of the regions of influence of competing shops. There is a similar problem for competing petrol stations, although competition between different brands manifests itself with many stations offering different brands, often close together.

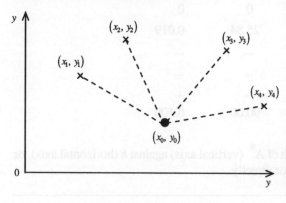

What are the important factors that determine the position of petrol stations?

In this case study, we will consider a related problem, namely that of the optimum location of the *depot* which provides the petrol to a specified number and fixed location of petrol stations. The situation is illustrated in this diagram for four petrol stations.

What are the important factors that determine the optimum position of the depot?

In the analysis below, we will describe a mathematical model and its solution for these problems, and then consider its viability.

In our model, we first assumed that:

(a) the depot is located at the point (x_0, y_0), using rectangular Cartesian coordinates (as shown in the previous figure);

(b) petrol stations are located at (x_1, y_1), (x_2, y_2), (x_3, y_3), ..., (x_n, y_n) ;

(c) the cost of delivery of petrol to the stations depends on

  (i) the demand at each station

  (ii) the distance of each station from the depot.

Activity 4.9

With the assumptions above, deduce a formula for the total cost of transporting petrol to the *n* petrol stations.

Using assumptions (i) and (ii) above, the simplest model is to take the cost of delivery to the *i*th stations as

$$c_i = \alpha w_i d_i$$

where α is a constant and

  w_i = demand (per week) for petrol at station *i*

  d_i = distance of petrol station *i* from depot.

Hence

$$d_i = \left\{(x_0 - x_i)^2 + (y_0 - y_i)^2\right\}^{\frac{1}{2}}$$

and the problem can be restated as: find the position (x_0, y_0) which minimizes the total delivery costs, that is

$$C = \alpha \sum_{i=1}^{n} w_i d_i$$

$$= \alpha \sum_{i=1}^{n} w_i \left\{(x_0 - x_i)^2 + (y_0 - y_i)^2\right\}^{\frac{1}{2}}$$

What variables does the cost depend on?

Unfortunately, C is dependent on two variables, x_0 and y_0.

To use calculus to find the minimum value of C requires solving

$$\frac{\partial C}{\partial x_0} = \frac{\partial C}{\partial y_0} = 0$$

(The notation $\dfrac{\partial C}{\partial x_0}$ means differentiate C with respect to x_0, keeping y_0 constant − it is called partial differentiation.)

Activity 4.10

Show that the conditions above require solving

$$\sum_{i=1}^{n} \frac{2(x_0 - x_i)w_i}{\left\{(x_0 - x_i)^2 + (y_0 - y_i)^2\right\}^{\frac{1}{2}}} = 0$$

and a second similar equation.

The two coupled equations in the activity above need to be solved, and, given the numerical values w_i, x_i, y_i ($i = 1, 2, ..., n$), a numerical technique could be used. Fortunately though, there is an easier and better way to proceed.

When the general case looks too complicated to cope with, it is often a good idea to try simpler (special) cases. For example, consider the special case of $n = 2$, as shown opposite. Clearly the depot must be somewhere on the straight line between the two stations.

The total cost is now given by

$$C = w_1 x + w_2(D - x)$$

(using the notation illustrated in the diagram above).

Thus $C = x(w_1 - w_2) + w_2 D$.

Where does C have a minimum value?

One missing factor in the equation for C is the constraint

$$0 \le x \le D$$

So plotting C against x will look something like the graph opposite.

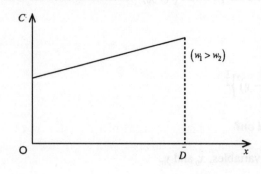

In this case ($w_1 > w_2$), the optimum location must be at petrol station 1. Similarly if $w_2 > w_1$, the depot should be at petrol station 2.

So the case $n = 2$ is straightforward, but may be of little help in finding the general solution. Let us look now at $n = 3$.

Activity 4.11

Consider the case of three petrol stations. Determine the optimum solutions in this special case when $w_1 = w_2 = w_3$.

The solution in the special case in the activity above is relatively straightforward. Provided the triangle formed by the three petrol stations has all its angles less than 120°, the depot will be located such that the lines to the petrol stations all make angles of 120°.

If the triangle has an angle greater than 120°, then the depot will be located at that position. These solutions are illustrated below.

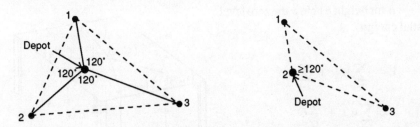

Similarly for equal weights for the case $n = 4$, the depot will be located at the intersection of the diagonals provided they do in fact intersect. If not, the quadrilateral formed by the positions of the petrol stations is non-convex, and the depot will be located at the 'internal' petrol station, as illustrated below on the right.

Unfortunately, whilst these special cases are of interest mathematically, they are of no real practical use, and, even more disappointingly, they do not appear to provide us with any help in finding the general solution.

Fortunately there is a way forward and it is based on a mechanical analog. Imagine that the demands at each of the petrol stations are replaced by weights.

Do you think that the centre of gravity of the system might be a suitable location?

In most practical cases the centre of gravity is probably a good approximation to the solution – but it is not the optimum solution as can be seen by looking at the $n = 2$ special case. Here the optimum solution is at one of the stations (the one with most demand) whilst the centre of gravity will be between the two stations (in the inverse ratio $w_1 : w_2$).

There is, though, a mechanical analog which does give the correct optimal solution. It is illustrated in the diagram below. A map of the region is pasted onto hardboard and holes made at each of the customer positions. Strings are passed through each of the holes and weights proportional to w_i attached to petrol station i, whilst the other ends of the string are attached to a smooth ring. If the length of the string through petrol station i is l_i, the potential energy of the system is

$$V = \sum_{1-i}^{n} w_i(l_i - d_i)$$

(since potential energy of a mass is given by $-mgh = -wh$ for height h below the zero level of potential energy).
Hence

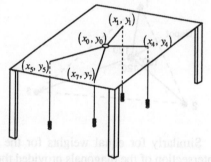

$$V = -\sum_{1=i}^{n} w_i l_i + \sum_{i=1}^{n} w_i d_i$$

$$= \sum_{1=i}^{n} w_i l_i + \frac{1}{\alpha} C$$

where C is the total cost.

So minimum potential energy will correspond to minimum C, since the other term in the equation $\left(\sum w_i l_i \right)$ is constant.

But, if released from rest (e.g. move the ring a small amount and let go), the system will take up its *equilibrium position* and this corresponds to *minimum potential energy*. If you are not familiar with this, consider a ball rolling over hills and valleys as illustrated below.

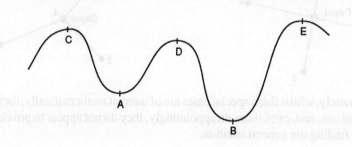

Points A, B, C, D and E are all equilibrium positions, but the potential energy of the ball has a minimum at A and B (and maximum at C, D and E). So *stable* equilibrium positions have minimum potential energy.

So, we have the solution – its position being given on the map as where the ring rests in equilibrium. The method is relatively crude but it is effective since the position can be seen. Also the effect of small variations in the demands, or the introduction of new petrol stations, can easily be seen.

What are the disadvantages of this method?

What assumptions made could be criticized?

How could the method be improved?

Two of the major drawbacks of the model are that it essentially assumes:

(i)   that there is a straight line road from each petrol station to the depot;

(ii)   that costs are linearly proportional to distance and demand.

Both these assumptions are difficult to modify in the model, so at best, the resulting solution is only a guide to the optimum position of a depot.

4.5   Road tanker design and loading

Road tankers
Materials such as petroleum products are transported in road tankers of the type shown in the diagram below.

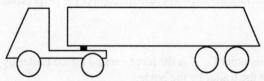

The unit consists of a *tractor,* and a *trailer* in which the material is stored.

The law specifies gross (total) weight limits for tankers of different types.

2-axle truck	16 tonnes
3-axle truck	24 tonnes
4-axle truck	31 tonnes
4-axle articulated	33 tonnes
5-axle truck	38 tonnes (38 000 kg)

The tankers are subject to additional safety requirements such as maximum permissible loads on axles. To check that legal safety limits have been met, the various components must be weighed separately. This is achieved using drive-on weighbridges.

The gross figures are calculated principally from environmental considerations (road wear, bridge loading limits, vibration of buildings, etc.) and not all trucks are designed to work to these limits. For example, axle design or tyre type may limit an axle loading to a value below the legal limit. Manufacturers are required to allot maximum loadings to axles and these loadings are stamped on a metal plate riveted to the vehicle. They are known as *plated weights*.

The trailer that contains the fuel is supported at the rear-end by a *bogie* and at the front by a *king pin*, by which it is attached to the tractor.

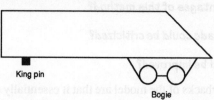

Consider an *empty* trailer PQRS whose cross-section is shown below.

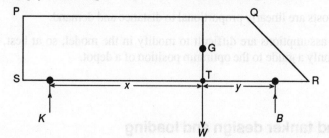

The centre of mass is that point through which the weight of the body (W) may be deemed to act. Since gravity acts vertically, we are concerned only with its horizontal location.

The trailer has been modelled as a combination of a rectangle and triangle. Hence the centre of mass will be located somewhere to the right of the perpendicular bisector of PQ.

Why is this so?

We have labelled this point G; K is the force exerted on the trailer by the king pin; B is the force exerted on the trailer by the bogie.

Since the vertical forces are in equilibrium, we have

$$K + B = W \qquad (1)$$

Also, the moments of the forces K and B about the point T are equal.

Explain why this is so.

Hence $\qquad\qquad xK = yB \qquad (2)$

Solving (1) and (2) gives

$$\boxed{K = \frac{y}{x+y}W \quad \text{and} \quad B = \frac{x}{x+y}W}$$

Since the centre of mass is nearer the bogie than the king pin, we have $y < x$ so that the force at the bogie is greater that the force at the king pin. K is called the *king pin tare weight* of the trailer and is an important measure in determining load safety.

Activity 4.12

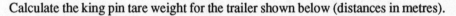

Where must the centre of mass must be located so that the king pin tare weight of the trailer is $\frac{1}{3}$ of the total weight?

Activity 4.13

Calculate the king pin tare weight for the trailer shown below (distances in metres).

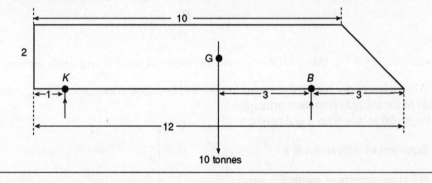

Single-compartment trailers

We consider now a trailer loaded with material that conforms to the shape of the trailer tank. This material is commonly fuel but could also be a substance such as grain. Again, we need to calculate the relative weights supported by the king pin and bogie.

Consider a single-compartment trailer (with load) as modelled below.

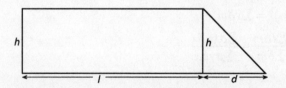

Suppose the material has a uniform density, ρ, and divide up the trailer as follows.

(1)   Horizontal component of the centre of mass of the rectangular cross-section is at $\frac{l}{2}$, by symmetry.

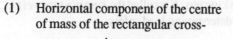

(2)

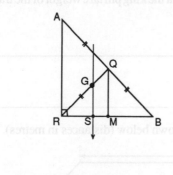

Centre of mass of a triangle is at the centroid ($\frac{2}{3}$ of the way along a median).

$$RG = \frac{2}{3}RQ \quad \text{where G is the centroid}$$

$$\frac{RS}{RM} = \frac{RG}{RQ}$$

$$\frac{RS}{\left(\frac{1}{2}a\right)} = \frac{2}{3}$$

Therefore $RS = \frac{1}{3}a$ (horizontal component of centre of mass of triangular section).

Alternatively, we may use calculus to find the horizontal coordinate of the centre of mass of the triangle from first principles.

Triangle OMN has base a and height h.

Equation of MN is $y = h\left(1 - \frac{x}{a}\right)$

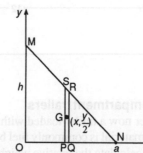

PQRS is an element of width δx, area approximately $y\delta x$ and mass $\delta m \approx \rho y \delta x$, where ρ is the uniform density of the material of the triangle.

Let $\bar{x}$ be the x-coordinate of the centre of mass of the triangle which has mass M.

Then $M \times \bar{x} =$ sum of moments of individual elements about the y axis (definition of centre of mass); giving,

$$\left(\sum \delta m\right)\bar{x} = \sum x \delta m$$

$$\bar{x} = \frac{\sum x \delta m}{\sum \delta m} = \frac{\sum \rho x y \delta x}{\sum \rho y \delta x}$$

Let $\delta x \to 0$

$$\bar{x} = \frac{\int_0^a xy\,dx}{\int_0^a y\,dx} = \frac{\int_0^a hx\left(1 - \frac{x}{a}\right)dx}{\int_0^a h\left(1 - \frac{x}{a}\right)dx}$$

Hence

$$\bar{x} = \left[\frac{1}{2}x^2 - \frac{x^3}{3a}\right]_0^a \Big/ \left[x - \frac{x^2}{2a}\right]_0^a$$

$$= \frac{1}{6}a^2 \Big/ \frac{1}{2}a$$

$$= \frac{a}{3}$$

Suppose a body of mass M is made up from components of mass m_1, m_2, m_3 ... and that each component has its own centre of mass with respective coordinates x_1, x_2, x_3 ...

Then the centre of mass of the compound body has coordinate $\bar{x}$ where

$$M\bar{x} = m_1x_1 + m_2x_2 + m_3x_3 + ...$$

That is, $\bar{x} = \dfrac{(m_1x_1 + m_2x_2 + m_3x_3 + ...)}{(m_1 + m_2 + m_3 + ...)}$

Another way of expressing this is to say that the moment about any point of the total mass (supposed concentrated at $\bar{x}$) is equal to the sum of the moments of the individual masses (supposed concentrated at their individual centres of mass).

We apply this principle in order to find the *centre of mass* of the *loaded* trailer.

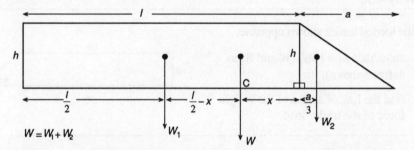

W_1 = weight of load in rectangular section which acts vertically through centre of rectangle

W_2 = weight of load in triangular section which acts in a line distance $\dfrac{a}{3}$ from the vertical side as shown in the previous section.

Let C be on the line of action of the total load. Then taking moments about C and using the above result, we have

$$W \times 0 = W_1\left(\frac{l}{2} - x\right) - W_2\left(x + \frac{a}{3}\right)$$

That is $x = \dfrac{3lW_1 - 2aW_2}{6(W_1 + W_2)}$

This locates the line of action of the total load weight.

Note also that the load weights may be expressed further in terms of the dimensions of the trailer. Let b be the (uniform) breadth of the trailer which is not visible in the cross-sectional diagram. Let ρ be the (uniform) density of the load material. Then

$$W_1 = \rho b(lh) = \rho blh$$

$$W_2 = \rho b\left(\frac{1}{2}ah\right) = \frac{1}{2}\rho bah$$

so $$W = \rho bh\left(l + \frac{1}{2}a\right)$$

Activity 4.14

Find the line of action of the weight force (location of horizontal component of centre of mass) for a trailer load of the dimensions given in the diagram.

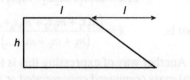

Activity 4.15

For the loaded trailer shown opposite,

(a)   show that $W_1 = 10W_2$ (W_1 and W_2 as defined above);

(b)   find the line of action of the weight force of the trailer load.

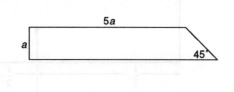

Activity 4.16

For each of the trailers shown below, find the line of action of the weight force of the trailer load.

(a)   (b)

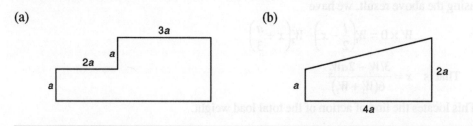

Activity 4.17

For the trailer shown below you are told that $l = \lambda a$ where λ is an integer.

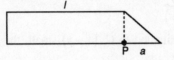

(a) Show that the line of action of the load weight is given by

$$x = \frac{a\,(3\lambda - 1)}{3\,(2\lambda + 1)}$$ where x is the distance from P to the line of action.

(b) Find the smallest integer value of λ for which $x > a$.

King pin and bogie loads

Earlier we considered king pin and bogie loads for empty trailers. We now investigate how to calculate the corresponding quantities for trailer loads. Consider the simple model of a loaded trailer shown below.

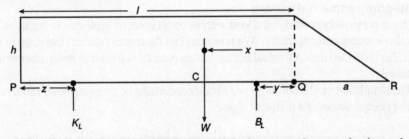

The bogie centreline, through which the weight is assumed to act, is situated at a distance y from Q. The king pin is situated z from P.

W is the total load weight and is comprised of $W_1 + W_2$ as described in the previous section.

For given values of W_1 and W_2 and the container dimensions the point C is fixed and is located at

$$x = \frac{3lW_1 - 2aW_2}{6W}$$

as found previously.

Let K_L and B_L be the forces exerted at the king pin and the bogie due to the load. We use the subscript L to indicate that these are load components and additional to forces produced by the empty trailer.

Moments about C give

$$W \times 0 = B_L(x - y) - K_L(l - x - z) \qquad (1)$$

Also $B_L + K_L = W$ (2)

Solving (1) and (2) for K_L gives

$$K_L = \frac{W(x-y)}{(l-y-z)}$$

The closer the bogie is situated to C, $(x-y)$ small, the smaller the load component supported by the king pin.

Activity 4.18

If $l = 3a$, find possible positions for the king pin and bogie so that the king pin load is equal to $\frac{1}{2}$ of the total weight.

Hint: $x = \dfrac{3lW_1 - 2aW_2}{6W}$

Multi-compartmental trailers

We have previously considered a load with its own centre of mass that is the same shape as a single-compartment trailer. We now consider the more common situation in which the trailer is divided into individual storage compartments – up to six or seven compartments are usual.

The simplified model below shows a trailer consisting of six equal compartments plus an end compartment of a different shape.

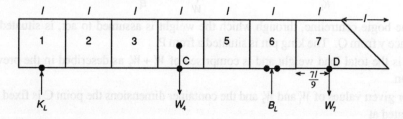

Now the load on the king pin consists of a contribution from each compartment.

$$K_L = K_{L1} + K_{L2} + \ ... \ + K_{L7}$$

where K_{L1} is the contribution from the load in compartment 1, etc. The same applies for bogie load B_L. For definiteness we assume that the kingpin is centred below compartment 1 and the bogie is centred below compartment 6.

We consider separately the contribution from each compartment. For example, here we will use compartment 4.

The weight of the load in compartment 4 (W_4) acts centrally through the centre of mass of compartment 4. Taking moments for this compartment about its centre, C, gives

$$W_4 \times 0 = B_{L4} \times 2l - K_{L4} \times 3l$$

so

$$3K_{L4} = 2B_{L4}$$

Also

$$W_4 = K_{L4} + B_{L4},$$

hence $K_{L4} = 0.4 W_4$

Consider now the contribution from compartment 7. The load in compartment 7 acts vertically through the centre of mass of compartment 7.

From Activity 4.14, the centre of mass of compartment 7 is located at D as shown in the diagram.

Taking moments for this compartment about D gives

$$W_7 \times 0 = B_{L7} \times \frac{23l}{18} + K_{L7} \times \frac{113l}{18}$$

since the distance of the bogie from D is $\frac{23l}{18}$, etc.

That is,

$$23 B_{L7} + 113 K_{L7} = 0$$

Also

$$B_{L7} + K_{L7} = W_7$$

so

$$K_{L7} = \frac{-23}{90} W_7 \approx -0.26 W_7$$

Note that compartment 7 contributes a negative component. Its centre of mass is not between the two points of support and counterbalances (lifts off in effect) some of the king pin load. Thus a loaded back compartment *reduces* king pin loading.

Activity 4.19

Calculate the loadings for the other compartments and check against the following:

$$K_{L1} = W_1 \qquad\qquad K_{L5} = 0.2 W_5$$
$$K_{L2} = 0.8 W_2 \qquad\quad K_{L6} = 0$$
$$K_{L3} = 0.6 W_3$$

The components 1, 0.8, 0.6, 0.4, 0.2, 0, 0.26 above are called *component factors*.

Total king pin load

In general, with a fully loaded truck (all compartments) the load on the king pin is given by

$$\text{Trailer tare} + K_{L1} + K_{L2} + L_{L3} + K_{L4} + K_{L5} + K_{L6} + K_{L7}$$

$$= K + \sum_{i=1}^{7} W_i \times (\text{factor for compartment } i)$$

Examples of king pin load calculation

i Compartment	W_i Product weight (tonnes)	K_{Li} Factor	King pin load (tonnes)
1	4.2	0.89	3.74
2	2.3	0.70	1.61
3	2.1	0.54	1.13
4	2.1	0.40	0.84
5	2.3	0.28	0.65
6	4.2	0.09	0.38
7	4.2	−0.12	−0.51
			7.84 tonnes
		Trailer tare king pin load	1.61
		Total king pin load	9.45 tonnes

(Source: Auto-motive Information Bulletin – Truck Analytical Loading and Discharge Procedure, Esso Petroleum Co. Ltd.)

Estimating the load

In practice, fuel is measured in litres so that conversion to units of weight (kg) is necessary to check that safety conditions are met.

This is achieved by using the simple formula

$$\text{litres / tonne} = \frac{1000}{\text{density}}$$

For example, if a product has a density (relative) of 0.84, then

$$\text{litres / tonne} = \frac{1000}{0.84} = 1190$$

Note: Petroleum products are sensitive to temperature change. Hence the value to be used is the actual density at the loading bay and not the standard value calculated at 15° C (often called specific gravity). Tables are available that enable corrections for temperature to be made.

Example

Loading temperature = 2° C
Specific gravity of product at 15° C = 0.843
Volume reduction factor (from tables) = 1.0106

So   Density at loading bay $= 0.843 \times 1.0106 = 0.852$

$$\text{litres/tonne} = \frac{1000}{0.852} = 1174$$

Without this adjustment, the calculation would have given

$$\text{litres/tonne} = \frac{1000}{0.843} = 1186$$

A 20 tonne load at 2° C is actually $20 \times 1174 = 23480$ l.

Using the standard (15° C) density gives $20 \times 1186 = 23720$ l.

Thus, without adjustments, the truck could have been overloaded by 240 litres (approximately 200 kg).

Activity 4.20

Using the product weight figures for the seven-compartment truck given in the table, and using the component factor calculated in Activity 4.19, calculate the product volume that could be loaded at 15° C and at 2° C, given the information in the preceding example.

Steps in load assessment

The planner's procedure for assessing a load consists of three distinct steps.

1. Ascertain maximum king pin (A) and gross loading (B) information for the truck/trailer combination.

2. Obtain product density information in terms of litres/tonne at the loading bay (see sub-section on *Estimating the load* on page 82).

3. Apply the appropriate calculations to ensure that king pin loadings and gross weights are not exceeded (see sub-section on *Total king pin load* on page 81).

For 1, the gross weight restriction is 32 imperial tons (32 300 kg approx.) for a truck on the road. The king pin load maximum is specified and is typically of the order of 10 tonnes.

Example

Referring to the table which follows, the load must be calculated and distributed among the compartments so that both A and B are satisfied.

The figures in the top two rows are given for a particular choice of vehicle. For a fixed product load the king pin load may be varied by altering the distribution of weights in the individual compartments.

	Plated weight (kg)	Tare weight (kg)	King pin load (kg)	
Tractor	17 000	6 820	10 180	max
		+	−	
Trailer	25 500	4 650	1 730	in tare condition
Total vehicle	32 500	11 470		
Max. product load on king pin			8 450	

Max product load on vehicle   32 500 − 11 470 = 21 030 kg

NB   • Weights always rounded to 10 kg due to weighbridge accuracy. In practice, there is a
        variation of ±20 kg between good weighbridges.
      • The law in the UK requires a maximum plated weight for the tractor as if it were on its
        own, the trailer as if it were on its own and the plated weight for the combination, to
        be shown.
      • The weight not to be exceeded (i.e. the maximum plated weight) for the tractor unit is
        made up of the tractor tare weight plus the trailer king pin tare weight plus the
        maximum product weight acting through the king pin.
      • The weight not to be exceeded (i.e. the maximum plated weight) for the trailer is
        made up of the trailer tare weight plus the product weight.
      • The weight not to be exceeded (i.e. the maximum plated weight(for the combination
        is made up of the tractor tare weight plus the trailer tare weight plus the product
        weight.

Activity 4.21

Assume that the specifications of a truck/trailer are as given in the example on page 83.
You are required to plan for the transport of a load of fuel whose density (15° C) has the
value 0.843. The temperature at the loading bay is 2° C (see sub-section on *Estimating the
load* on page 82). Complete the following table so that the product load is as large as
possible without exceeding the king pin load.

Your solution will consist of the entries in the *Actual volume carried* column for each
of the six compartments provided for in the following table. The other entries in the table
will involve checking the entries to ensure that legal limits are met.

Compartm't number	Max. volume	Actual volume carried	Product density	Actual weight in compartm't	Proportion of compartm't weight on king pin	Actual compartm't load on king pin
		a	+ b	= c	× d	= e
1	5 000				0.95	
2	5 000				0.69	
3	5 000				0.47	
4	5 000				0.26	
5	5 000				0.05	
					sub total	
6	5 000				−0.16	
Totals						
Max legal totals				20 451		7593

Write calculated values of legal limits here (from previous chart)

Discharging the load

Because a loaded back compartment on a trailer has a counterbalancing effect, it actually lifts weights off the king pin (see section on *Multi-compartment trailers,* page 80). Therefore by discharging this compartment first, it is possible to overload the king pin by removing the counterbalancing influence. Similarly, discharging compartment 1 (front) lifts a lot of weight off the drive axle. If compartment 1 only is discharged, the truck does not handle well, so this option should be avoided also.

It is normal to load the densest product into the rear compartment to reduce the weight on the king pin. The weight on the king pin tends to be the critical weight on UK vehicles.

Activity 4.22

The following diagrams show trailer tanks loaded in different ways. In each case the load consists of two consignments or drops, both having the same density.

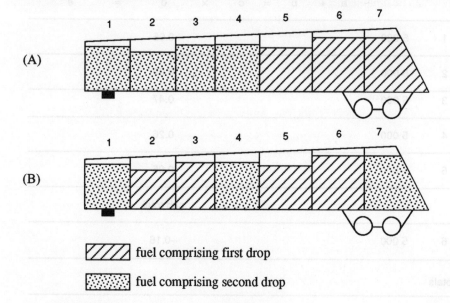

(a)   Explain which of (A) or (B) is the better arrangement.

(b)   It has been claimed that when a load consists of two or more drops of products with equal densities, one drop should be distributed near the ends of the tank and the other(s) in the middle compartments. Argue for or against this claim.

Activity 4.23

Suppose that the load you planned in Activity 4.21 comprises two drops representing (approximately) 40% and 60% of the total respectively.

(a)   Decide which compartments should be used for the respective drops.

(b)   After making the first drop, recalculate the king pin loadings.

4.6 Related data

Esso publishes many useful resources (including videos) for educational use, including qualitative information of relevance to case studies in this chapter. A catalogue of resources can be obtained from

> Esso Information Service
> PO Box 695
> Sudbury
> Suffolk
> CO10 6YM

More general enquiries concerning the company should be addressed to

> Public Affairs Department
> Esso UK plc
> Esso House
> Victoria Street
> London
> SW1E 5JW

This section includes a variety of data with which you will be able to test hypotheses, look for correlations, and analyze and present the trends in data.

Some sample questions are given below.

Questions

1. Is there any correlation between the company's turnover and profit (after tax)?

(Table 4.6.1)

2. Describe the trends in US average wellhead prices and pump prices.

(Table 4.6.2)

3. What has been the significant change in oil trade movement over the past 10 years?

(Table 4.6.4)

4. For 1993, seek the correlation between crude imports, product imports with crude exports or product exports. (Table 4.6.5)

5. What relationship is there between estimated oil reserves and oil production?

(Table 4.6.6)

4.6.1 Esso UK plc: results 1987–1993

	1987	1988	1989	1990	1991	1992	1993
Gross turnover	£5.8bn	£5.3bn	£5.4bn	£5.7bn	£6.2bn	£6.5bn	£6.6bn
Profit (after tax)	£527m	£351m	£299m	£303m	£310m	£305m	£388m
Tax*	£0.8bn	£0.6bn	£0.3bn	£0.2bn	£0.2bn	£0.3bn	£0.3bn
Return on capital	18%	11%	11%	9.9%	8.5%	8%	8.2%
Asset base	£3.6bn	£3.7bn	£4.2bn	£4.6bn	£5.2bn	£5.5bn	£5.5bn
Investment	£379m	£553m	£672m	£790m	£600m	£666m	£643m

* Corporation and Petroleum Revenue tax only (excise duty and VAT excluded)

4.6.2 Price history: crude oil and leaded petrol

Year	U.S. average wellhead price $/bbl	Leaded petrol pump price $/gal	Year	U.S. average wellhead price $/bbl	Leaded petrol pump price $/gal
1960	2.88	0.311	1976	8.14	0.590
1961	2.89	0.308	1977	8.57	0.622
1962	2.90	0.306	1978	8.96	0.626
1963	2.89	0.304	1979	12.51	0.857
1964	2.88	0.304	1980	21.59	1.191
1965	2.86	0.312	1981	31.77	1.311
1966	2.88	0.321	1982	28.52	1.222
1967	2.92	0.332	1983	26.19	1.157
1968	2.94	0.337	1984	25.88	1.129
1969	3.09	0.348	1985	24.09	1.115
1970	3.18	0.357	1986	12.51	0.857
1971	3.39	0.364	1987	15.40	0.897
1972	3.39	0.361	1988	12.58	0.899
1973	3.89	0.388	1989	15.86	0.998
1974	6.74	0.532	1990	20.03	1.149
1975	7.56	0.567			

4.6.3 Pump prices 1950–1993

Petrol retailing in the United Kingdom is a highly competitive business, and each site operator sets the prices for that site. There is no single, or uniform price, and the figures quoted below are therefore typical only.

Date*	Pump price 4-star	Tax		Taxes as % pump price	Pump price** excl. taxes
1950	£0.15	7.5p		50.0%	7.5p
1960	£0.24	12.5p		53.6%	11.2p
1970	£0.31	22.5p		73.0%	8.5p
1980	£1.39	45.5p +	18p VAT	45.7%	75.5p
1983	£1.79	74.1p +	23p VAT	45.7%	75.5p
1984	£1.86	78.0p +	24p VAT	54.8%	84.0p
1985	£2.05	81.5p +	21p VAT	53.0%	96.5p
1986	£1.65	88.1p +	21p VAT	66.4%	55.4p
1987	£1.73	88.1p +	22.5p VAT	64.2%	55.4p
1988	£1.78	92.9p +	23.2 VAT	65.3%	61.7p
1989	£1.93	92.9p +	25.2p VAT	61.2%	74.9p
1990	£2.00	102.2p +	26p VAT	64.1%	71.8p
1991	£2.31	117.5p +	34p VAT	65.5%	79.5p
1992	£2.36	126.3p +	35.1p VAT	68.4%	74.4p
1993	£2.36	139.0p +	35.1p VAT	73.8%	61.9p

* Dates are mid-year   ** Prices are in pence per gallon

4.6.4 Oil trade movements 1983–1993 Thousand barrels daily

	1983	1984	1985	1986	1987	1988	1989	1990	1991	1992	1993
Imports											
USA	4990	5380	5065	6045	6245	7240	8019	8026	7791	7888	8527
OECD Europe	9006	8938	8733	9242	8238	9438	9702	9747	10118	10252	10333
Japan	4145	4305	4045	4140	4125	4412	4549	4802	4925	5306	5307
Rest of world*	6555	6470	6645	7220	6315	7111	8320	8866	9504	9540	10321
Total world	24696	25093	24488	26647	24923	28201	30590	31441	32338	32986	34488
Exports											
USA	740	720	780	765	745	845	817	889	1000	918	959
Canada	545	655	685	675	630	890	944	955	1111	1101	1215
Latin America	4065	34100	3565	3600	3035	3246	3511	3754	3421	3843	3825
Middle East	10355	9845	9340	10880	10315	11842	13389	14212	13829	15453	16456
North Africa	2180	2290	2415	2515	2435	2598	2415	2604	2781	2849	2685
West Africa	1425	1670	1765	1980	1880	2022	2319	2248	2500	2679	2676
Asia & Australasia**	1465	1655	1600	1610	1410	1380	1758	1582	1699	1847	1982
Rest of world	925	950	1050	1220	1230	1242	1641	1938	3579	1842	2180
Non-OECD, Europe & China	2996	3208	3288	3402	3243	4136	3796	3259	2418	2454	2510
Total world	24696	25093	24488	26647	24923	28210	30590	31441	32338	32986	34488

* Includes unidentified trade   ** Excludes China and Japan

4.6.5 Oil imports and exports 1993

	Million tonnes				Thousand barrels daily			
	Crude imports	Product imports	Crude exports	Product exports	Crude imports	Product imports	Crude exports	Product exports
USA	335.2	85.9	5.3	40.8	6732	1796	106	853
Canada	29.3	8.0	45.1	14.8	688	167	906	309
Latin America	55.7	14.2	142.9	45.7	1119	297	2870	955
OECD Europe	429.9	81.3	41.0	34.3	8633	1700	823	717
Middle East	4.5	3.9	729.2	86.7	90	82	14644	1812
North Africa	4.6	5.1	103.1	29.4	92	107	2070	615
West Africa	2.1	6.5	129.8	3.3	42	136	2607	69
East & Southern Africa	18.6	6.6	–	0.5	374	138	–	10
South Asia	30.0	19.4	0.2	2.2	602	406	4	46
Other Asia	162.4	54.9	51.0	31.9	3261	1148	1024	667
Japan	218.9	43.6	0.1	5.9	4396	911	2	123
Australasia	20.1	2.2	7.4	4.4	404	46	149	92
Non-OECD Europe & China	35.6	24.3	87.4	36.1	715	508	1755	755
Unidentified*	–	–	4.4	19.9	–	–	88	416
Total world	1346.9	355.9	1346.9	355.9	27049	7440	27049	7440

* Includes changes in the quantity of oil in transit, movements not otherwise shown,
  unidentified military use, etc.

Note: Intra-area movements (for example, between countries in OECD Europe) are excluded.

4.6.6 Reserves and production

| | Estimated proved reserves | | | | Oil production | | | |
| | Jan 1, 1992 | | Jan 1, 1991 | | | | | |
COUNTRY	Oil	Gas	Oil	Gas	Producing wells 31.12.90	Estimated 1991 *	Change from 1990 (%)	Actual 1990 *
Western Europe								
Austria	84,984	388	81,443	403	1,242	25.9	−19.4	21.7
Denmark	755,000	4,059	799,435	4,488	98	140.4	−16.3	120.7
France	170,836	1,296	184,766	1,324	647	58.5	−4.4	61.2
Germany	449,000	8,760	425,000	12,400	2,100	74.3	−3.6	77.1
Greece	41,000	44	30,000	47	13	15.3	−6.7	16.4
Ireland	...	1,700	...	1,700	...	...	...	...
Italy	692,153	11,370	693,503	11,624	222	89.5	−4.8	94.0
Netherl'nds	144,650	69,570	157,200	60,900	187	66.3	−5.7	70.3
Norway	7,609,400	60,670	7,609,412	60,674	308	1,864.0	15.0	1,620.5
Spain	21,389	740	20,000	780	44	21.3	31.5	16.2
Turkey	540,450	705	650,000	1,150	684	86.7	20.1	72.2
U.K.	3,994,310	19,247	3,825,000	19,775	880	1,779.3	−3.8	1849.7
Total	14,503,172	178,549	14,475,759	175,265	6,425	4,221.5	5.0	4,020.0

Asia-Pacific								
Total	44,073,216		50,242,057		63,524		3.2	6,257.5
		299,288		301,091		6,456.6		

Eastern Europe and USSR								
Total	58,773,500		58,855,000		150,545		−10.1	11,702.9
		1,766,358		1,619,000		10,524.3		

Middle East								
Total	661,570,779		662,597,860		7,314		−3.3	16,784.5
		1,319,118		1,324,265		16,248.5		

* thousands of barrels per day

CHAPTER FIVE

Tesco

5.1 Location of stores

Over the past few years there has been a phenomenal increase in the number of out-of-town shopping centres. Those involving Tesco sometimes consist of one large store or, alternatively, a large Tesco store may be one of a complex of shops. Planners of such centres need to estimate the likely success of such a venture. Information concerning

- the size of the shopping centre
- how much parking space will be needed
- ease of access to the centre
- areas from which shoppers will be attracted

will be required.

Activity 5.1

List other important questions or issues that need to be addressed in the development of a new shopping centre.

Some of the larger shopping centres like Brent Cross in North London have attracted shoppers from a very wide area, but this has caused significant parking problems at peak times (e.g. on Saturday mornings). A complete analysis of the likely profit from opening a new out-of-town Tesco store is very complex, so here we will just provide some insight into one aspect of the problem. This relates to the area of influence of a particular shop. There is little profit in building two Tesco out-of-town stores within a mile of each other!

What criteria need to be considered in deciding how near two stores could be to each other?

The mathematical analysis is based on defining the variable

> T = number of journeys made (in a specified time interval) to a particular store.

Suppose a residential zone is a distance d away from the store.

How will T vary with d ?

Clearly as d decreases, T will increase, but at what rate? The usual model taken to describe this situation is an *inverse square* model of the form

$$T = \frac{F}{d^2}$$

Here F is a measure of the attractiveness of the store.

What factors will F depend on?

It is difficult to put a precise value on F, but it is possible to estimate the attractiveness of one store when compared with another. This will depend on size, parking, associated petrol station, bus connections, other facilities, etc.

Activity 5.2

Consider two neighbouring Tesco out-of-town stores that you are familiar with. Ask friends to judge their relative merits (e.g. shop A is twice as attractive as shop B, etc.) and see how consistent the judgements are.

For planning purposes, it would be useful to estimate the area of influence of two neighbouring stores. That is, we wish to determine the boundary of choice between one store and another.

On one side of the boundary shoppers will tend to go to one centre; on the other side they will normally go to the other centre. The situation is shown in the diagram opposite, when the two stores are at $A(0, 0)$ and $B(b, 0)$, using a coordinate system with origin at A.

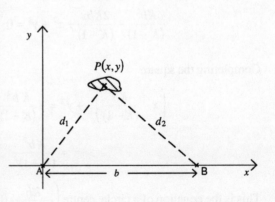

Consider a residential zone centred at P which has coordinates (x, y).

What is the condition for points on the boundary choice?

The number of journeys to store A is given by

$$T_A = \frac{F_1}{d_1^{\,2}} = \frac{F_1}{x^2 + y^2}$$

when F_1 is the attractiveness of store A; whereas the number of journeys to store B is given by

$$T_B = \frac{F_2}{d_2^{\,2}} = \frac{F_2}{(b-x)^2 + y^2}$$

Points on the boundary of choice will make an equal number of journeys to store A and store B; that is

$$\frac{F_1}{x^2 + y^2} = \frac{F_2}{(b-x)^2 + y^2}$$

This can be simplified by writing $K = \left(\dfrac{F_1}{F_2}\right)$, so that

$$K\{(b-x)^2 + y^2\} = x^2 + y^2$$

What type of equation does this give?

The equation can be rewritten as

$$Kb^2 - 2Kbx + (K-1)x^2 + (K-1)y^2 = 0$$

or

$$\frac{Kb^2}{(K-1)} - \frac{2Kbx}{(K-1)} + x^2 + y^2 = 0$$

Completing the square

$$\left(x - \frac{Kb}{(K-1)}\right)^2 + y^2 = -\frac{Kb^2}{(K-1)} + \frac{K^2 b^2}{(K-1)^2}$$

$$= \frac{Kb^2}{(K-1)^2}$$

This is the equation of a circle, centre $\left(\dfrac{Kb}{(K-1)}, 0\right)$ and radius $\dfrac{b\sqrt{K}}{(K-1)}$.

The solution for $K = 3$ is illustrated below.

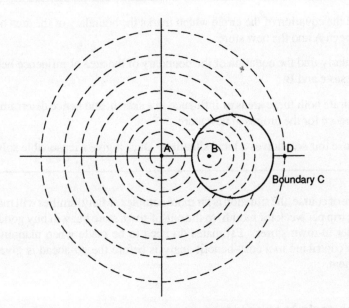

Of course, $K = 3$ implies that we have noted one centre three times as attractive as the other. This is quite extreme, but it serves to illustrate the important aspects.

The circles round each centre are curves on which T is constant. Note that for all points outside the boundary C, the store preferred is A. For example, points to the far east of the boundary (e.g. point D) would prefer to bypass the store at B and to shop at A, which is further away in distance, but has much more attractive facilities.

Activity 5.3

Illustrate the solutions for $K = 2$ and $K = 4$. What happens if $K = 1$?

This type of analysis can be extended to planning a new out-of-town store midway *between* two established town centre stores.

As an example, suppose the proposed new store, with every conceivable facility, is noted six times more attractive than the town centre store A, and three times more attractive than the town centre store B. The situation is shown in this diagram.

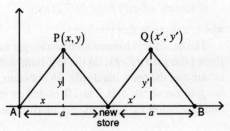

Activity 5.4

(a) Find the equation of the circle which marks the boundary of the area of influence between A and the new store.

(b) Similarly find the equation of the boundary of the area of influence between the new store and B.

(c) Illustrate both these areas of influence in a sketch, and hence determine the area of influence for the proposed new store.

(d) Impose this solution as a local situation. Does it give a reasonable solution?

In practice of course, the situation is far more complex. Many families will make one large shopping trip per week (or month) to the out-of-town store but will buy goods daily from the smaller in-town stores. Estimates do need to be made when planning, and these estimates contribute to a cost–benefit analysis before the go-ahead is given for such a development.

5.2 Bar code technology

Bar codes are almost universal today, being used in just about every industry. All but the smallest Tesco stores now use this technology, not just for checkout pricing, but also for stock control.

Bar codes were first suggested for automation in grocery stores in 1932 in the thesis of a Harvard Business School student but it was not until the 1950s that the idea of a scanner installed at checkouts was conceived. It took another two decades for a combination of technology advancement and economic pressure to bring about the commercial use of bar codes and optical readers in retail trading.

In 1973 the UPC (Universal Product Code) was adopted as a standard and in 1976 a variation known as the EAN (European Article Number) was also standardized.

Other types of bar codes, 'Code 3 of 9' and 'Interleaved Two of Five (ITF)' have also been developed for other purposes.

An example of an eight-digit EAN symbol is shown below. This appears on

Cadbury's Dairy Milk BUTTONS

and is unique to this product.

The bar code is formed with two longer vertical lines (the *guard lines*) on the left-hand side, the centre and the right-hand side. In between, there are *four* numbers, 5, 0, 2 and 0, coded on the left and *four* numbers, 1, 2, 3 and 5, coded on the right.

left-hand guard   centre guard   right-hand guard

Left-hand five

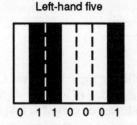

0 1 1 0 0 0 1

In fact, each digit is represented by a seven-module system. For example, a left-hand side 5 is represented by the bar code shown on the left.

Each digit has *two white* and *two black* strips of varying thicknesses but following the rules that:

(i) the first 'module' must be white;
(ii) the last 'module' must be black;
(iii) there are, in total, either three or five black modules.

A convenient way of representing each number is given by using 0 (white), 1 (black) giving 0110001 for 5 (as shown).

Activity 5.5

With the rules above, design all possible codes for left-hand numbers.

How many such codes exist?

Activity 5.6

Find as many eight-digit EAN symbols as you can (do not use 13-digit EAN symbols) from products.

Look very carefully at the *left*-hand representation of numbers and see if you can match up the designs found in Activity 5.5 to the digits 0–9.

In fact, the full set of codes is given on page 101; left-hand numbers are represented by the number set A.

The representation of numbers on the right-hand side is related to the left-hand side representation, but is *not* the same.

Why is a different representation needed for left and right sides?

Right-hand five

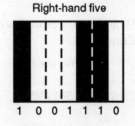

1 0 0 1 1 1 0

The right-hand side representation of digit 5 is shown opposite.

What is the relationship between the left- and the right-hand sides' representations of numbers?

To see the answer to this, look carefully at the codes.

Left-hand five	0	1	1	0	0	0	1
Right-hand five	1	0	0	1	1	1	0

One code is the *dual* of the other.

Right-hand representation of digits is given on page 101 by the number set C.

One important concept which most bar codes incorporate is that of a *check digit*. To avoid errors made in optical reading (particularly when codes are on curved or rough surfaces), the *final digit*, which is the check digit, is always based on the previous ones. When a bar code is read, the computer will verify that the check digit is correct before accepting the number.

The check digit is determined by the condition that

$$3 \times (1\text{st} + 3\text{rd} + 5\text{th} + 7\text{th digit}) + (2\text{nd} + 4\text{th} + 6\text{th digit}) + \text{check digit}$$

is divisible by 10.

For example, for the number

$$5020\ 1235$$

we require

$$3 \times (5 + 2 + 1 + 3) + (0 + 0 + 2) + 5$$

to be divided by 10.

The total gives

$$3 \times 11 + 2 + 5 = 40$$

and so this obeys the rule for check digits.

Activity 5.7

Check the labels with eight-digit EANs to see that they are correctly labelled.

Will one error in the optical reading always be detected?

Activity 5.8

Here are three examples of eight-digit EANs. Are they correct? If not, can you correct them?

   (a)   0004 6121

   (b)   5019 9588

   (c)   5099 9713

We now move on to 13-digit EAN symbols. These are found on many products and particularly on Tesco own brands.

You can see, on the right, an example from a pack of Tesco jelly. The first digit, which as you can see is not represented directly in the code, together with the second digit, indicates the country in which the article number was allocated, e.g. 50 represents UK, 31 represents France, etc. The next five digits are issued to a particular manufacturer, and the next five identify the product. The final number is again the check digit.

The check digit works in a similar way.

$$3 \times (0 + 8 + 7 + 2 + 2 + 2) + (5 + 1 + 3 + 4 + 3 + 7) + 4$$

$$= 63 + 23 + 4$$

$$= 90 \quad \text{which is divisible by 10.}$$

Activity 5.9

Find some 13-digit EANs. Check the accuracy of the check digit for each one.

All the six right-hand digits are coded using number set C (see page 101) but the six left-hand digits are coded using a combination of number sets A and B, according to the first digit. For example, if the first digit is 5, then the next six digits are coded using the number sets   A B B A A B   for each digit in turn.

Activity 5.10

Using three As and three Bs, how many different possible combinations exist for the coding?

In fact, the first digit 0 uses the code A A A A A A whereas all other first digits are coded using three As and three Bs as indicated in the table on page 101.

Exercise 5A

1.   Find the check digits for each of the following 13-digit codes.

(i)

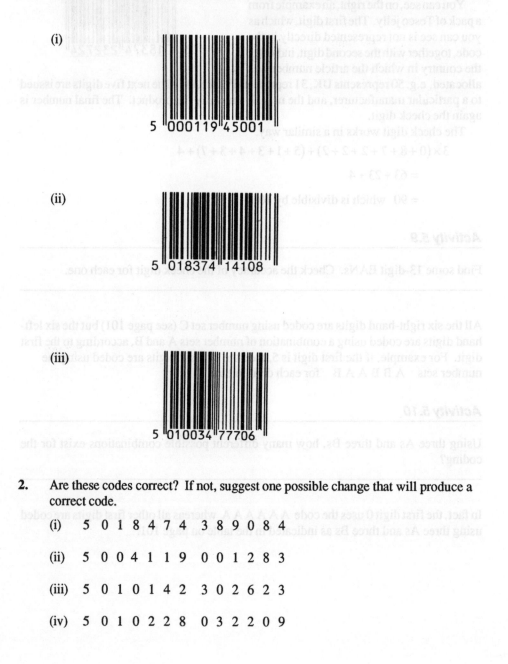

5 000119 45001

(ii)

5 018374 14108

(iii)

5 010034 77706

2.   Are these codes correct? If not, suggest one possible change that will produce a correct code.

(i)   5 0 1 8 4 7 4   3 8 9 0 8 4

(ii)   5 0 0 4 1 1 9   0 0 1 2 8 3

(iii) 5 0 1 0 1 4 2   3 0 2 6 2 3

(iv) 5 0 1 0 2 2 8   0 3 2 2 0 9

Number sets

Value of digit	Number set A (odd)	Number set B (even)	Number set C (even)
0			
1			
2			
3			
4			
5			
6			
7			
8			
9			

Coding system for left-most digit of EAN 13

Value of digits	Number sets for coding left-hand numbers
0	A A A A A A
1	A A B A B B
2	A A B B A B
3	A A B B B A
4	A B A A B B
5	A B B A A B
6	A B B B A A
7	A B A B A B
8	A B A B B A
9	A B B A B A

5.3 Stock control

The use of bar code technology not only speeds up service to customers at the checkout but also provides a computerized stock control system.

Daily (or even part-daily) information on sales gives an almost instant facility for stock ordering, and the time scale between ordering stock and delivery (often straight onto the shelf) is remarkably small. It is no longer necessary for shops to hold a considerable amount of stock – this is taken care of at the warehouses. In fact Tesco has a range of distribution centres, conveniently located for quick and efficient transportation to local stores. There are for example, six regional distribution centres, located at

Crick, Welham Green, Weybridge, Westbury and Middlewich

which provide the bulk of 'medium fast' ranges of ambient (room temperature) products.

Activity 5.11

Check, on a map of the UK, the location of these five distribution centres. Are their locations convenient for distribution to stores in their particular region?

In fact, Tesco also has two more distribution centres for just wines and spirits (at Harlow and Minworth), a national centre for all hardware and textile lines (at Milton Keynes), and eight 'composite' distribution centres. These are so called since they carry any combination of frozen and fresh-chilled grocery products. (There are temperature-controlled compartments in the delivery vehicles.)

Each of these composite distribution centres is about the size of Wembley Stadium; they are located in

Chepstow
Didcot
Doncaster
Harlow
Hinckley
Livingston
Maidstone
Middleton

Activity 5.12

Check the location of the composite distribution centres. Are there any areas of the UK not adequately serviced by these centres?

The optimum location of depots is an interesting (mathematical) problem but here we will look more clearly at the way in which stock is controlled in these large warehouses. Orders for a particular item will clearly be subject to random fluctuations about a trend, and the pattern of stock level with time will follow a pattern of the form shown below.

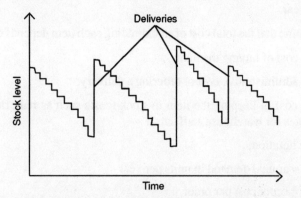

To minimize costs, the warehouse manager will not want to keep too much stock in the warehouse, but it is important that the stock level does not decrease to zero!

What are the cost implications of having too much stock of a particular item?

The relationships shown in the diagram above are rather too complicated to model exactly.

What simple assumption can be made to the model?

Our first simple assumption is that there is a demand spread evenly over the year so that an amount Q is regularly ordered in time period T. We also assume that it is important not to run out of stock completely at any time, so that there is always an amount, s, in stock. This gives the situation illustrated in the diagram below, where Q is known as the lot size, T, as the reorder period, and s as the safety stock.

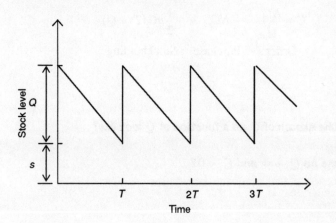

Looking at this simplified graph, we see that we have assumed that the stock level is repeatedly dropping from $s + Q$ to s in a linear manner. Thus the *average* stock held throughout the year is $s + \frac{1}{2}Q$. Also, with a reorder period of T years there will be $\frac{1}{T}$ deliveries per year.

Finally, we assume that the total cost of stockholding each item depends on three factors:

1.   the cost of buying the item;

2.   the administrative cost of ordering a delivery;

3.   the cost of keeping the item in stock (costs such as rent, rates, wages for warehouse staff, etc.).

Introducing the notation,

N = annual demand in units per year

C_1 = order cost per order

C_2 = purchase cost per item

a = stockholding cost (as fraction of price) per item,

enables us to write the total annual cost as

$$C = \frac{1}{T}C_1 + NC_2 + \left(s + \frac{Q}{2}\right)aC_2$$

Order   Purchase   Stockholding

We also note that T and Q are related via $\left(\frac{1}{T}\right)Q = N$ so that

$$C = \frac{NC_1}{Q} + NC_2 + \frac{1}{2}aC_2(2s + Q)$$

Order   Purchase   Stockholding

What does the sketch of C as a function of Q look like?

What happens as $Q \to \infty$ and $Q \to 0$?

Whatever the values of the parameters N, C_1, C_2, a and s, the shape of the curve will also take the form shown below.

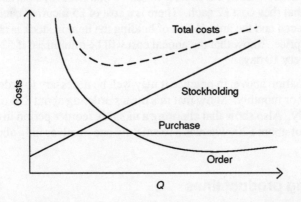

The key characteristic is that the 'total cost' graph has a *minimum* for positive Q. That is, there is a value of Q, the stock order size, which gives rise to a minimum value for the total costs.

How can you find the critical value of Q?

One method is to differentiate C with respect to Q, and set

$$\frac{dC}{dQ} = 0$$

This gives

$$\frac{dC}{dQ} = \frac{NC_1}{Q^2} + 0 + \frac{1}{2}aC_2$$

$$= 0 \quad \text{when} \, Q = \left(\frac{2NC_1}{aC_2}\right)^{\frac{1}{2}}$$

This value of Q, at which the total annual cost will be a minimum, is known as the *economic batch quantity* (EBQ) or economic order quantity. The corresponding value of T, the reorder time, is

$$T = \frac{Q}{N} = \left(\frac{2C_1}{aNC_2}\right)^{\frac{1}{2}}$$

These EBQ formulae form the basis for nearly all stockholding and purchasing policies.

Exercise 5B

1. Suppose that the warehouse expects to sell 10 000 items of a particular type this year, and that they cost £2 each. There is a cost of £5 incurred each time an order is placed and the annual cost of holding the item in stock is about 18 per cent of its price. Show that the annual cost will be minimized if 521 items are ordered every 19 days.

2. For the situation above, in practice it may well be necessary to order the item fortnightly or monthly. Show that this means ordering about 834 or 417 items respectively. Also show that choosing a monthly reorder period involves an extra cost of about £20, whereas a fortnightly one involves only about £5 excess.

5.4 Sampling product lines

Tesco is very concerned about the quality of its products, and all products are regularly sampled to ensure consistency. Required standards of consistency are sent to all suppliers, and all aspects of product storage and distribution are monitored to ensure that the quality is never compromised during the distribution chain to the point of sale.

The monitoring of quality is often based on taking samples and testing for quality, usually referred to as *conforming* or *non-conforming* (i.e. defective – but this is no longer an acceptable word). A sampling plan is devised which gives the criteria for acceptance of the complete batch. The following case study illustrates this.

A pottery firm produces coffee mugs, decorated with the faces of famous sports personalities. The following two sampling plans have been suggested.

Method A (single sample plan) – select 20 mugs from the batch at random and accept the batch if there are two or less non-conforming, otherwise reject the batch.

Method B (double sample plan) – select 10 mugs from the batch at random and accept the batch if there are no non-conforming mugs, reject if there are two or more non-conforming, otherwise select another 10 mugs at random. When the second sample is drawn, count the number of non-conforming mugs in the combined sample of 20 and accept the batch if the number of non-conforming items is two or less, reject otherwise.

Discuss the advantages or disadvantages of Method B compared with Method A.

The effectiveness of either method can be analyzed by first assuming that the proportion of non-conforming items is p. We can then find the probability of accepting the batch for each method, and evaluate this for a range of values of p.

So for Method A, if the probability of obtaining a non-conforming item is p, and the sample size is 20, then (using the binomial distribution*) the probability of obtaining 2, 1 or 0 non-conforming items is

$$p(2 \text{ non-conforming}) = {}^{20}C_2\, p^2(1-p)^{18}$$

$$p(1 \text{ non-conforming}) = {}^{20}C_1\, p(1-p)^{19}$$

$$p(0 \text{ non-conforming}) = (1-p)^{20}$$

So, adding up these probabilities gives

$$p(2 \text{ or less non-conforming}) = 190p^2(1-p)^{18} + 20p(1-p)^{19} + (1-p)^{20}$$

$$= (1-p)^{18}\left(190p^2 + 20p(1-p) + (1-p)^2\right)$$

$$= (1-p)^{18}\left(1 + 18p + 171p^2\right)$$

Hence the probability of accepting the batch is given by

$$\boxed{(1-p)^{18}\left(1 + 18p + 171p^2\right)}$$

For Method B, the probability of accepting on the first sample is $(1-p)^{10}$.

A second sample is taken if a non-conforming item is found. This has probability

$${}^{10}C_1\, p(1-p)^9 = 10p(1-p)^9$$

The batch is then accepted if only 0 or 1 more non-conforming items are found in the second sample. This has probability

$$(1-p)^{10} + 10p(1-p)^9$$

and so the combined probability is given by

$$10p(1-p)^9\left((1-p)^{10} + 10p(1-p)^9\right)$$

So, for Method B, the probability of acceptance is

$$(1-p)^{10} + 10p(1-p)^9\left((1-p)^{10} + 10p(1-p)^9\right)$$

$$= (1-p)^{10} + 10p(1-p)^{18}(1-p+10p)$$

$$= (1-p)^{10} + 10p(1-p)^{18}(1+9p)$$

* The binomial distribution gives the probability of x successes in independent trials as

$$P(x) = \binom{n}{x}p^x(1-p)^{n-x}, x = 0,1,...,n$$

where p is the underlying probability of success.

This gives the probability of accepting the batch as

$$(1-p)^{10}\left[1+10p(1-p)^8(1+9p)\right]$$

Now to see how these formulae can be used.

Activity 5.13

Evaluate the formula found above for a variety of values of p.

e.g. $p = 0.01, 0.05, 0.1, 0.2, 0.5$, etc.

From the activity above you can find the probabilities of acceptance, e.g. when $p = 0.01$ and $p = 0.5$.

	Probability	
Method	0.01	0.2
A	0.999	0.206
B	0.995	0.208

There is surprisingly little difference between the methods, so Method B is probably the better of the two to use, since it probably requires less sampling.

Activity 5.14

Consider the following sampling methods.

Method A (single sample plan) – select 10 mugs from the batch at random and accept the batch if there are two or less non-conforming, otherwise reject the batch.

Method B (double sample plan) – select five mugs from the batch at random and accept the batch if none are non-conforming, reject if there are two or more non-conforming, otherwise select another five mugs at random. When the second sample is drawn, count the number non-conforming in the combined sample of 10 and accept the batch if the number of non-conforming mugs is two or less, otherwise reject the batch.

(a)   If the proportion of non-conforming mugs in a batch is p, find, in terms of p, for each method in turn, the probability that the batch will be accepted.

(b)   Evaluate both the above probabilities for $p = 0.2$ and $p = 0.5$.

(c)   Hence, or otherwise, decide which of these two plans is more appropriate, and why.

Activity 5.15

Design your own sampling method and analysis in a similar way to that above.

5.5 Packaging

Packaging products is an important aspect of the grocery trade. At Tesco, there is a design department dedicated to producing packaging that meets legal requirements, such as

- Product description
- Size and position of weight statement, price and 'sell-by' features
- Ingredients
- Country of origin
- Tesco name and address
- Sizes and positioning of the above

and also Tesco's own requirements

- Tesco logo
- Clear typeface, especially for storing and cooking instructions
- Imperial/metric weights
- Clear visible product where desirable
- Distinctive identity to each in a range of products
- In-store display factors
- Prices and 'sell-by' methods
- Suppliers' code
- Bar code considerations
- Customer information on nutrition and environmental factors.

In this case study, we will consider an earlier problem, that of optimum wrapping of a product. As a simple example, consider the problem of wrapping with clear plastic film (which is then heat sealed) magazines of size

$$20 \text{ cm} \times 20 \text{ cm} \times 4 \text{ mm}$$

$$30 \text{ cm} \times 40 \text{ cm} \times 4 \text{ mm}$$

What is the minimum amount of plastic film required?

For a first solution, it seems sensible to ignore the thickness of the magazine (4 mm) and just consider wrapping an 'area'. We can then add on 2 cm on each dimension to take account of the thickness and overlap needed for sealing. Three methods will be considered.

Method A – 'square-on'

There are two possible solutions as illustrated below for the general case of covering an area of a cm by b cm.

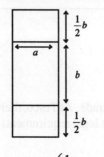

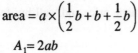

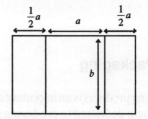

$$\text{area} = a \times \left(\frac{1}{2}b + b + \frac{1}{2}b \right) \qquad\qquad \text{area} = b \times \left(\frac{1}{2}a + a + \frac{1}{2}a \right)$$

$$A_1 = 2ab \qquad\qquad\qquad\qquad A_2 = 2ab$$

Adding on 2 cm for the overlap gives

$$A_1 = (a+2)(2b+2) \qquad\qquad A_2 = (b+2)(2a+2)$$

i.e. $A_1 = 2ab + 2a + 4b + 4 \qquad\qquad A_2 = 2ab + 4a + 2b + 4$

Explain why $A_1 \neq A_2$.

What happens to the formula for A_1 and A_2 when $a = b$?

Since

$$A_1 - A_2 = (2ab + 2a + 4b + 4) - (2ab + 4a + 2b + 4)$$

$$= 2(b-a)$$

then if $b > a$, $A_1 - A_2 > 0$, and A_2 will require a smaller area.

Activity 5.16

Find the areas A_1 and A_2, for the two cases

(i)   20 cm × 20 cm   (ii)   40 cm × 30 cm

Method B – '45°'

This method is illustrated opposite. By symmetry

$$WB = a\sin 45° = \frac{a}{\sqrt{2}} \text{ and } BX = b\sin 45° = \frac{b}{\sqrt{2}}.$$

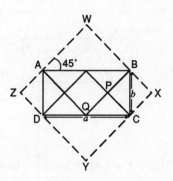

So

$$WX = WB + BX$$

$$= \frac{a}{\sqrt{2}} + \frac{b}{\sqrt{2}}$$

$$= \frac{1}{\sqrt{2}}(a+b)$$

Again allowing 2 cm for overlap and sealing, the required area is given by

$$A_3 = \left(\frac{1}{\sqrt{2}}(a+b)+2\right)^2 = \frac{1}{2}\left(a+b+2\sqrt{2}\right)^2$$

Activity 5.17

Evaluate A_3 for

    (i)   20 cm × 20 cm   (ii)   40 cm × 30 cm

Method C – 'minimum angle'

It is clear from the previous method that the area of overlap can be reduced by turning the wrapping through some angle other than 45° (except, of course, when the article being wrapped is square, when there is no overlap).

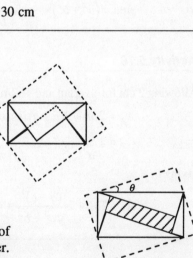

However, it is also clear that if the angle is too small the top surface will not be covered, as in the diagram opposite.

As the angle θ is increased, the two long sides of the uncovered area will rotate and get closer together. (Remember that all four triangles will change shape so that the edges will still just meet at the corners; the size of the piece of film must also be increased.)

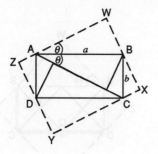

The previously uncovered area will just be covered when θ is the same as the angle between the diagonal and the top edge of the item.

i.e. $\theta = B\hat{A}C = \tan^{-1}\dfrac{b}{a}$ (from triangle ABC)

$WX = YZ = a\sin\theta + b\cos\theta$
$XY = WZ = a\cos\theta + b\sin\theta$.

Area of film $= (a\sin\theta + b\cos\theta)(a\cos\theta + \sin\theta)$
$= a^2\sin\theta\cos\theta + ab\sin^2\theta + ab\cos^2\theta + b^2\sin\theta\cos\theta$
$= (a^2 + b^2)\sin\theta\cos\theta + ab$

Since

$$\tan\theta = \frac{a}{b}, \quad \sin\theta = \frac{b}{\sqrt{a^2 + b^2}}, \quad \cos\theta = \frac{a}{\sqrt{a^2 + b^2}}$$

so $\text{area} = (a^2 + b^2) \times \dfrac{b}{\sqrt{a^2 + b^2}} \times \dfrac{a}{\sqrt{a^2 + b^2}} + ab = 2ab$

Activity 5.18

Allowing 2 cm for overlap and sealing, show that the required area is given by

$$A_4 = 2ab + 2(a + b)(\sin\theta + \cos\theta) + 4$$

Since $\tan\theta = \dfrac{b}{a}, \quad \sin\theta = \dfrac{b}{\sqrt{a^2 + b^2}}, \quad \cos\theta = \dfrac{a}{\sqrt{a^2 + b^2}},$

giving

$$A_4 = 2ab + 2\frac{(a + b)^2}{\sqrt{a^2 + b^2}} + 4$$

Activity 5.19

Evaluate A_4 for

    (i)   20 cm × 20 cm   (ii)   40 cm × 30 cm

Which of the methods described above gives the smallest value for A?

Try evaluating A_1, A_2, A_3 and A_4 for other dimensions of the area.

Does the 2 cm overlap affect the calculations and results?

Of course, in practice, it is unlikely that the sort of analysis presented above will be undertaken. Appearance, ease of handling, etc., are probably far more important in the success of a product than a small saving in packaging – but when we can, we should also use the method that minimizes the wastage.

5.6 Related data

Further information concerning Tesco can be obtained from

> Tesco Distribution Division Ltd
> Caswell Road
> Brackmills Industrial Estate
> Northampton
> NN4 7TW

This section contains tables of data which is presented for you to analyze, hypothesize or complete. The data includes sections on the finances of Tesco, food prices, household food consumption and the Retail Price Index. Some sample questions for you to answer are given below.

Questions

1. What are the main points of interest from the recent company data?   (Table 5.6.1)

2. Is there any correlation between profit and number of employees or number of stores or sales area?   (Table 5.6.1)

3. Comparing price increases from 1950 to 1990, what items do you think are good value now?   (Table 5.6.2)

4. Describe the main trends of the food price history in Table 5.6.3.

5. Rank the consumption values given in Table 5.6.4 for 1986 and 1992. Do these correlate well?

5.6.1 Company data: five-year record for Tesco

Year ended February	1989	1990	1991	1992[1]	1993
Financial statistics £m					
Turnover excluding VAT	4,717.7	5,401.9	6,346.3	7,097.4	7,581.5
Operating profit	276.5	334.0	420.0	503.3	577.2
Operating margin[2]	5.9%	6.2%	6.6%	7.1%	7.6%
Interest receivable less payable	2.4	9.8	19.1	65.5	31.5
Employee profit sharing	(13.6)	(17.2)	(22.0)	(23.8)	(25.5)
Profit before net (loss)/surplus on sale of properties	265.3	326.6	417.1	545.0	583.2
Net margin[2]	5.6%	6.0%	6.6%	7.7%	7.7%
Net (loss)/surplus on sale of properties	10.7	35.0	19.1	0.5	(2.3)
Profit before taxation	276.0	361.6	436.2	545.5	580.9
Taxation	(89.7)	(107.8)	(133.5)	(149.9)	(163.3)
Profit after taxation	186.3	253.8	302.7	395.6	417.6
Earnings per share[3] – pence	11.98	15.87	18.37	20.43	21.45
Fully diluted earnings per share (excl net (loss)/surplus on sale of properties)[3] – pence	10.98	13.35	16.60	19.95	20.97
Dividends per share[3] – pence	3.40	4.17	5.25	6.30	7.10
Net worth – £m[4]	1,031.3	1,254.1	2,159.9	2,447.0	2,752.9
Return on shareholders' funds[5]	27.9%	28.6%	29.4%	23.7%	22.4%
Return on capital employed[6]	22.1%	22.5%	21.2%	19.3%	18.3%
Net assets per share – pence	67	80	112	126	141
Productivity £					
Turnover per employee	89,449	99,400	106,044	119,246	130,612
Profit per employee	5,243	6,146	7,018	8,456	9,944
Wages per employee	8,695	10,009	10,579	12,250	13,172
Weekly sales per sq ft	12.30	13.61	15.06	15.47	15.69

Notes:

1 53-week period

2 Based upon turnover exclusive of VAT

3 Adjusted in respect of 1991 rights issue

4 Total shareholders' funds at the year end

5 Profit before net (loss) /surplus on sale of properties divided by weighted average shareholders' funds

6 Profit before net (loss) /surplus on sale of properties and interest dividend by average capital employed

Company data (continued)

Year ended February	1989	1990	1991	1992[1]	1993
Retail statistics					
Market share in food & drink shops	8.0%	8.4%	9.0%	9.4%	9.7%
Number of stores	374	379	384	396	412
Total sales area – '000 sq ft	7,986	8,442	8,956	9,661	10,352
Sales area opened in year – '000 sq ft	514	718	798	889	859
Average store size (sales area) – sq ft	21,400	22,300	23,300	24,400	25,100
Average sales area of stores opened in year – sq ft	32,100	31,300	38,100	35,500	34,400
Full-time equivalent employees	52,742	54,345	59,846	59,519	58,046
Share price – pence					
Highest	169	216	246	296	293
Lowest	129	151	194.5	207	197.5
Year end	153	196	246	271	235

5.6.2 Food price history – Table 1

Commodity	Unit of sale	|| Average prices (pence)	1950	1960	1970	1980	1990
White bread	1¾lb		5.5	11.5	21.0	32.3	50.0
Flour (white)	3lb		9.75	21.4	24.0	37.1	55.0
Butter	1lb		24.0	38.9	44.7	81.0	125.0
Cheese	1lb		14.0	33.8	46.1	98.4	151.0
Eggs (large)	each		4.5	5.9	4.5	6.0	10.0
Milk	pint		5.0	8.0	12.0	17.0	32.0
Tea	1lb		40.0	78.7	81.2	112.8	243.3
Coffee (pure roasted)	1lb		–	86.0	140.0	236.1	295.1
Sugar (white granulated)	2lb		5.0	16.1	18.3	37.0	59.0
Fish (fresh cod)	1lb		18.1	33.0	52.7	108.7	282.0
Beef (sirloin)	1lb		26.0	65.0	111.0	218.9	281.3
Bacon (back, smoked)	1lb		37.0	58.7	83.6	121.3	196.0
Beer (bottled)	pint		–	22.4	34.6	40.5	97.0
Potatoes	7lb		10.7	19.1	4.2	5.4	14.0
Oranges	1lb		9.0	11.3	16.5	22.9	41.0

5.6.3 Food price history – Table 2

Average prices paid for household foods, 1984–1992

	1984	1985	1986	1987	1988	1989	1990	1991	1992
Liquid wholemilk	21.41	22.57	23.61	24.79	25.78	27.69	20.87	31.33	31.99
Yoghurt	59.56	64.86	69.46	75.56	80.58	87.17	96.59	102.81	108.57
Cream	155.27	152.83	152.55	146.45	145.74	148.85	162.11	142.81	135.10
Natural cheese	119.56	126.34	128.88	132.50	146.41	155.46	163.36	166.55	181.75
Beef and veal	157.49	161.35	164.87	164.56	179.20	198.46	196.15	203.39	201.68
Pork	119.70	120.63	122.12	127.05	127.20	142.39	158.22	155.37	171.33
Liver	74.22	72.92	78.25	85.29	88.10	92.85	100.41	97.37	102.36
Sausages (uncooked pork)	79.45	81.97	85.31	87.38	91.74	97.38	108.28	112.67	113.24
Sausages (uncooked beef)	72.71	72.98	74.77	75.74	79.87	86.17	93.06	92.89	92.97
Fish white (filleted fresh)	126.56	134.48	154.05	171.11	188.57	195.59	220.12	252.35	258.44
Shellfish	287.72	282.44	305.05	347.32	344.62	391.18	371.44	367.91	360.48
Frozen fish	111.70	119.09	125.66	133.98	141.38	144.65	166.19	175.12	171.69
Eggs	6.78	6.88	6.97	7.38	7.48	8.03	9.21	9.22	9.25
Butter	86.40	84.89	91.44	91.21	96.07	107.81	109.79	108.84	111.93
Margarine	43.36	47.28	42.37	40.65	42.67	45.64	49.96	55.02	57.56
Sugar	22.11	22.15	22.43	23.89	26.07	27.46	29.40	31.61	30.32
Old potatoes	11.54	5.86	7.01	9.42	8.81	8.89	13.33	12.14	12.62
Carrots	16.86	17.23	17.93	20.31	21.84	21.46	25.70	27.27	22.88
Mushrooms	97.68	101.38	101.65	104.57	109.47	111.21	117.93	119.89	124.25
Frozen peas	39.15	39.28	40.65	37.51	44.54	46.24	48.37	48.55	49.37
Apples	30.59	30.24	33.55	34.25	35.55	36.79	44.99	51.67	49.52
Bananas	37.00	41.17	43.72	45.10	45.74	46.65	49.69	51.43	46.38

Prices – pence per lb, except for the following: per pint of milk, yoghurt, cream, per egg.

5.6.4 Household consumption of individual foods: 1986–1992

	(ounces per person per week)						
	1986	*1987*	*1988*	*1989*	*1990*	*1991*	*1992*
Milk and cream	4.01	3.94	3.89	3.84	3.71	3.66	3.82
Cheese	4.15	4.07	4.13	4.07	4.00	4.11	4.00
Meat	23.78	24.34	24.08	23.72	22.95	22.62	23.45
Fish	5.13	5.06	5.04	5.18	5.06	4.85	4.97
Eggs	2.91	2.79	2.58	2.21	2.12	2.18	2.03
Fats	10.48	10.01	9.85	9.48	9.00	8.75	8.65
Sugar and preserves	9.97	9.30	8.75	8.18	7.69	7.60	7.07
Potatoes	36.52	35.63	34.54	33.70	33.53	32.55	30.74
Fresh green veg	8.81	7.81	8.47	8.34	8.07	8.25	8.09
Other fresh veg	14.99	15.07	15.27	15.40	14.67	15.13	15.69
Processed vegetables	19.56	19.46	19.42	19.06	18.62	19.16	19.97
Fresh fruit	18.70	18.72	19.81	20.13	20.32	20.46	21.15
Fruit products	10.44	10.75	10.87	10.95	10.22	11.98	10.93
Bread	30.75	30.55	30.25	29.41	28.06	26.49	26.61
Cereals	54.88	54.79	54.06	53.25	51.74	51.24	51.55
Beverages	2.76	2.70	2.66	2.59	2.46	2.50	2.35

5.6.5 Retail Price Index

The retail price index (RPI) is an important barometer of the economy of the UK and shows the impact of inflation on family budgets. The inflation figures are eagerly watched by the media, wage-bargainers, business people and politicians. The index affects us all. It can affect your tax allowances, your savings, state benefits and pensions since these and many other payments are often updated using the index.

The RPI is not strictly a *cost of living* index, a concept which means different things to different people. To many it may suggest the changing costs of basic essentials, but in practice it would be very difficult to agree on a definition of 'essentials'. The index simply gives an indication of what we would need to spend in order to purchase the same things we chose to buy in an earlier period, irrespective of whether particular products are 'needed' or 'good for you'. For example, some people buy cigarettes, so these are included in the index. However the calculations reflect the average shopping basket and so take account of the fact that a majority of people do not buy tobacco at all.

We spend more on some things than others and we would expect, for example, that an increase in mortgage interest payments would have a considerably greater impact on the RPI than an increase of similar proportion in the price of tea. The components of the index are therefore carefully 'weighted' to ensure that the index reflects the importance of the various contents of the average shopping basket and the amounts we spend in different regions of the country and in different types of shops.

Index of retail prices

Annual averages

16 January 1962 = 100			15 January 1974 = 100			13 January 1987 = 100		
	All items	Food		All items	Food		All items	Food
1966	116.5	115.6	1974	108.5	106.1	1987	101.9	101.4
1967	119.4	118.5	1975	134.8	133.3	1988	106.9	105.7
1968	125.0	123.2	1976	157.1	159.9	1989	115.2	111.9
1969	131.8	131.0	1977	182.0	190.3	1990	126.1	120.8
1970	140.2	140.1	1978	197.1	203.8	1991	133.5	128.6
1971	153.4	155.6	1979	223.5	228.3			
1972	164.3	169.4	1980	263.7	255.9			
1973	179.4	194.9	1981	295.0	277.5			
1974	208.2	230.0	1982	320.4	299.3			
			1983	335.1	308.8			
			1984	251.8	326.1			
			1985	373.2	336.3			
			1986	385.9	347.3			

CHAPTER SIX

Mathematics in the 1990s

6.1 Introduction

In the preceding chapters you have seen many applications of mathematics. Sometimes a mathematical model has been used to solve a particular problem; at other times, mathematical models have been used to illustrate particular phenomena or to make predictions. All of these applications, however, are characterized by the process shown in diagrammatic form below.

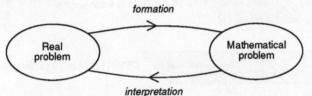

The 'real' problem, i.e. the problem from real life, has to be translated into a mathematical one. This is done essentially by forming a mathematical model. The main variables have to be defined and, if appropriate, the equations relating them need to be specified. Sometimes this process is complex and not at all straightforward. At other times it is quite clear what to do.

Having formulated the model, usually there is a resulting mathematical problem. Again it could be a straightforward evaluation of an expression, but often it requires the solving of equations or the optimizing of a function. This is where your mathematical skills are needed.

The final stage is to translate that solution back in terms of the real world. This may be uncomplicated but it also may need careful consideration of the solution to the real problem. Does it make sense? Does it correspond with any practical evidence? This is a crucial step in the modelling cycle if the mathematical model is to have any real value.

Activity 6.1

A single-lane (in each direction) straight tunnel, one mile long, has been built to help increase traffic flow in a city centre. Given that the Highway Code gives advice about minimum separation distances between vehicles (Rule 57 states that you should never get closer to the vehicle in front than the overall stopping distances, i.e. no closer than the combined thinking and braking distance relevant to the speed), what advice would you give the tunnel operators about recommended speeds in the tunnel?

For information, the Highway Code recommendations are given below.

Shortest stopping distances

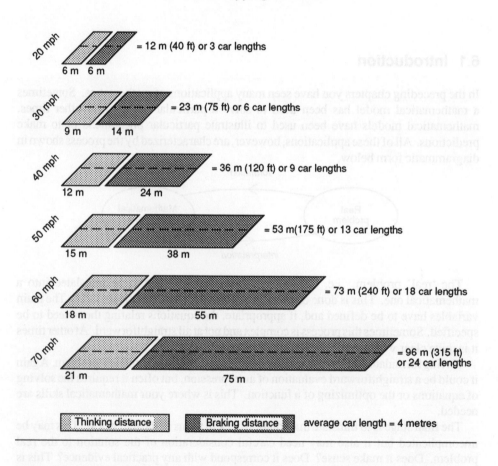

20 mph — 6 m, 6 m — = 12 m (40 ft) or 3 car lengths

30 mph — 9 m, 14 m — = 23 m (75 ft) or 6 car lengths

40 mph — 12 m, 24 m — = 36 m (120 ft) or 9 car lengths

50 mph — 15 m, 38 m — = 53 m (175 ft) or 13 car lengths

60 mph — 18 m, 55 m — = 73 m (240 ft) or 18 car lengths

70 mph — 21 m, 75 m — = 96 m (315 ft) or 24 car lengths

Thinking distance   Braking distance   average car length = 4 metres

This section is completed with two recent applications, both from the commercial world but very different in outlook.

6.2 Channel Tunnel

At the time of going to press, the Channel Tunnel has just opened for passenger through-trains between London and Paris or Brussels. These are specially designed trains which need to operate on three different types of power generators, namely the 'third rail' electric lines south of London and the two systems of overhead electricity which operate on the continent. The trains (called Eurostar and operated by EPS) will be gradually phased in over the next two years as more become available. This case study explores the design of timetables that can be operated with different numbers of trains.

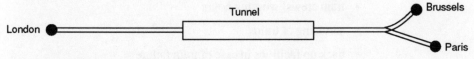

Current journey times are

$$\text{London} - \text{Paris} \quad : \quad 3 \text{ hours}$$
$$\text{London} - \text{Brussels} \quad : \quad 3\tfrac{1}{4} \text{ hours}$$

When a train arrives at a terminus, at least one hour is needed before it is ready to depart.

If, for example, *four* trains were initially available, a possible timetable based on local time (continental Europe is one hour *ahead* of UK time), is shown below.

Waterloo International	08.23	15.53
Paris Nord	12.23	19.53
Paris Nord	08.07	17.09
Waterloo International	10.13	19.09
Waterloo International	10.23	16.23
Brussels Midi/Zuid	14.38	20.47
Brussels Midi/Zuid	08.28	18.26
Waterloo International	10.43	20.43

This timetable uses four trains but it does not make best use of them. For example, one train is idle from its arrival in Paris at 12.23 hours until its departure for London at 17.09 hours.

Activity 6.2

Redesign the timetable given above so that better use is made of the four trains. Take into account the following practical conditions:

- each train needs one hour at its destination station before it is ready to leave;

- no train must leave London, Paris or Brussels before 07.00 hours, local time;

- all trains must arrive at their final destination by 22.00 hours local time, at the latest;

- there must be at least ten minutes separation time between any two trains travelling in the same direction on the same route.

Of course, the real problem is even more complex as consideration must be given to factors such as

- train crews' working hours

- servicing of trains

- back up facilities in case of train failure

and many other factors! An important concept which you should have incorporated with your solution to Activity 6.2 is the provision of trains at suitable times for passengers and with suitable intervals between services. In 1996, though, there should be a regular service of trains to/from Paris and Brussels.

Activity 6.3

A regular hourly service from London to Paris and return, and a two-hourly service from London to Brussels and return, are scheduled for 1996.

What is the minimum number of trains required to provide this level of service (assuming all the conditions given in Activity 6.2 still apply)?

Each Eurostar train normally consists of

- power car at one end

- sixteen seating coaches (either first class or standard class)

- two buffet coaches

- power car at other end

The seating coaches are either first class (with 39 seats) or standard class (with 60 seats).

The order of magnitude of revenue for each class is about £90 for first class, £60 for standard class, per journey.

Why is using only first class coaches not a very sensible idea?

Activity 6.4

Investigate the income from different combinations of first class coaches and standard class coaches. Suggest a sensible combination of each type.

6.3 National Lottery

Towards the end of 1994 a national lottery was started in the UK, although it should be noted that such events have been commonplace in many countries throughout the world for many years. Most lotteries are designed in a similar way. The participant buys a ticket for a specified fee, and on this ticket are printed their choice of r numbers (r is specified) out of a total n numbers. The jackpot is won if this choice is the same set of numbers as that given by the lottery draw. Usually repeats are not allowed, and the order of the numbers is irrelevant.

For example, the UK lottery requires a choice of 6 numbers out of 49 (1 to 49). This is called a '6/49 lottery'.

Other examples are

California	6/53
Virginia	6/44
New York	6/54

Of course, everyone wants to win the jackpot – but what are the chances?

Is the UK lottery a good risk, or would it be better to leave your money in Premium Bonds, or risk it gambling on horses, or the football pools?

We will not be able to give a complete, precise answer but you can at least quantify the odds. In fact, the UK agency, Camelot, who run the lottery, give the odds in their publicity. The information is shown in the table below.

Winning selections	Odds	Expected prize
Jackpot: match 6 main numbers	1 in 13,983,816	£2,000,000 (win or share)
Match 5 main numbers plus the bonus number	1 in 2,330,636	£100,000
Match 5 main numbers	1 in 55,492	£1,500
Match 4 main numbers	1 in 1,033	£65
Match 3 main numbers	1 in 57	Guaranteed £10

The overall odds of winning a prize are 1 in 54

The value of prizes, except the £10 fixed prize, is entirely dependent on the number of winners and tickets sold.

Our first task is to see if we agree with the stated odds. For example, for the jackpot in which all six numbers must be correctly matched, we can start by considering the number of possible outcomes.

There are 49 ways of choosing the first number. Having chosen it, there are 48 contenders for the second number, etc. In this way, there are

$$49 \times 48 \times 47 \times 46 \times 45 \times 44$$

different ways of choosing the six numbers.

But this takes into account the order of the numbers; that is, the choice, for example, 1, 2, 3, 4, 5, 6 also occurs as 2, 1, 3, 4, 5, 6.

How many ways are there of permuting six different numbers?

You can use a similar argument to answer this. There are six ways of choosing the first number, five ways of choosing the second number, and so on. This gives

$$6 \times 5 \times 4 \times 3 \times 2 \times 1$$

permutations of six different numbers.

Hence the number of different combinations possible is

$$\frac{49 \times 48 \times 47 \times 46 \times 45 \times 44}{6 \times 5 \times 4 \times 3 \times 2 \times 1}$$

Is each possible combination equally likely to occur?

Since each combination is equally likely to occur, the odds of winning the jackpot are

$$1 \text{ in } \frac{49 \times 48 \times 47 \times 46 \times 45 \times 44}{6 \times 5 \times 4 \times 3 \times 2 \times 1}$$

$$\Rightarrow \quad 1 \text{ in } 13\ 983\ 816$$

which confirms the stated result on the publicity. In the UK, each ticket costs £1.

How much would it cost to buy a ticket for every combination?

So if the lottery jackpot exceeded about £14 million, you could be sure to win (provided you were the only winner!). But, in the UK, if no one wins the jackpot, the money is added to the jackpot for the next week and this continues for two further games. Consequently, *unless* the public buy an excessive number of tickets, and provided no one wins the jackpot for three consecutive weeks, there is no chance of guaranteeing a win!

In fact, worldwide a number of syndicates have conspired (quite legally) to win. For example, in the Virginian lottery after the jackpot had reached $27 million (each ticket cost $1), an Australian, Peter Mandel, with the help of 20 workers, bought a ticket for every combination.

How many combinations are there for the Virginian 6/44 lottery?

Peter Mandel managed to buy only 90 per cent of the tickets (as two of his workers did not fulfil their tasks) but this was enough for him to win and net a tidy profit!

Those of you familiar with combinatorics will know the general result; that is, the number of combinations of r different objects from a total of n different objects is given by

$$^nC_r = \frac{n!}{r!(n-r)!} = \frac{n(n-1)(n-2) \dots (n-r+1)}{r(r-1)(r-2) \dots 2 \times 1}$$

You can use your scientific calculator to evaluate these coefficients directly, since it will have a '^nC_r' button.

Check $^{49}C_6$ and $^{44}C_6$ on your calculator.

So we have confirmed the first prize. The second prize is more if you match five of the six numbers plus the bonus number (which is a seventh number drawn).

As before, there are $^{49}C_6$ possible combinations. Of these, the number of ways of choosing five out of the six numbers is $^6C_5$ and the number of ways of choosing the bonus number is 1. Hence the probability of winning the second prize is

$$\frac{^6C_5 \times 1}{^{49}C_6} = = \frac{6}{13\,983\,816} = \frac{1}{2\,330\,636}$$

or 1 in 2 330 636.

Activity 6.5

Using the same sort of analysis

(a) show that the probability of winning the third prize, i.e. the probability of matching five main numbers but not the bonus number, is given by $\dfrac{^6C_5 \times {}^{42}C_1}{^{49}C_6}$;

(b) find a formula for winning the fourth and fifth prizes.

By now you should have realized that the general formula for the probability of matching x out of the r winning numbers, chosen for n different numbers, is given by $\dfrac{^rC_x \, {}^{n-r}C_{r-x}}{^nC_r}$

Verify that this formula gives the correct results in (a) and (b) in the Activity above.

Activity 6.6

Show that the overall odds of winning a prize are 1 in 54.

Armed with all this information, you can now design a lottery. Be sure, though, to check that if your design were to be put into practice, the odds would be such that a sufficiently large profit would be generated for the organizers whilst still giving participants some hope of winning a substantial prize.

Example

A lottery organized to raise funds for a college is designed as a 2/6 lottery with tickets costing £1 each, and with the following prizes:

£4 for both numbers matching

£2 for just one number matching

Is the college likely to make a profit?

Solution

In such a simple lottery it is easy to list all the possible tickets, i.e.

1, 2	2, 3	3, 4	4, 5	5, 6
1, 3	2, 4	3, 5	4, 6	
1, 4	2, 5	3, 6		
1, 5	2, 6			
1, 6				

There are $5 + 4 + 3 + 2 + 1 = 15$ different tickets.

Thus the probability of winning the first prize is $\dfrac{1}{15}$ and for the second is $\dfrac{8}{15}$. Hence the organizer's expected profits per ticket, in £s, are

$$1 - 4 \times \frac{1}{15} - 2 \times \frac{8}{15} \; = \; -\frac{1}{3}$$

So, on average, the college will lose £$\dfrac{1}{3}$ per ticket, i.e. £100 for every 300 tickets sold!

Can you suggest different prize money to ensure that a profit is made for the college?

Activity 6.7

Design your own simple lottery, e.g. 3/6, 2/8, 3/8, etc.

Work out the associated probabilities and decide on the appropriate prize money which could be offered.

6.4 Postscript

Some of the ways in which mathematics is used in the real world have been shown through various case studies and examples. We hope that you will have appreciated the variety of ways in which maths is used to help solve practical problems.

As technology advances rapidly, the need for mathematical skills is undoubtedly as important as ever. The mathematical skills needed may change as the technology changes, but what remains constant is the need to find efficient and effective solutions to real world problems.

Answers

Chapter 1 INTRODUCTION

Exercise 1A

1. (a) −11.45°C (b) −4.71°C
   (c) −18.18°C
2. 1016, 181.7 cm
3. Easter Sunday: 7 April 1996,
   30 March 1997, 12 April 1998

Exercise 1B

2. Competitor number 6
3. $L = 25 - 2W$
4. 49 (with 30 m waste);
   40 (with 20 m waste);
   10 (with 8 m^2 waste)
6. $\dfrac{R}{R_e} = 3 \times 2^n + 4$;

   $n = -\infty$ for Mercury

Chapter 2 BRITISH STEEL

Exercise 2A

1. sphere
2. height $= \sqrt{3} \times$ side length
3. $1 : 2$

Exercise 2B

1. $\left(1 - \dfrac{\pi}{2\sqrt{3}}\right) \times 100 \approx 9.31\%$
2. 10

Exercise 2C

1. 0.0027%
2. $\left(\dfrac{2R\sin^{-1}\left(\dfrac{l}{2R}\right)}{l} - 1\right) \times 100\%$

   where $R = \dfrac{h}{2} + \dfrac{l^2}{8h}$

Exercise 2D

2. (a) 48 (b) 86 (c) 78

Exercise 2E

1. thickness $= l\left(1 - \dfrac{r}{100}\right)^n$;

   speed $= \dfrac{v}{\left(1 - \dfrac{r}{100}\right)^n}$
2. $n = 3$

Exercise 2F

1. (a) 21 646 m   (b) 4329 m

2. $L = \dfrac{\pi(D^2 - d^2)}{4\omega}$

Chapter 3 THE POST OFFICE

Exercise 3A

1. (i) $S = 3416$

   • • • • • • • • • • • •

   (ii) $S = 75$

   • • • • • • • • • • • •

   (iii) $S = 1497$

   • • • • • • • • • • • • •

2. (i) 3HU   (ii) 0DU   (iii) 4FZ

3. 4095, so 1364 and 2729 divide
   all S values into three equal groups

Exercise 3B

The total distances are unique but there
are many routes possible.

1. 1120   e.g. C B A D B E D E C
2. 54 000
3. 905
4. 40 000

Chapter 5 TESCO

Exercise 5A

1. (i) 2   (ii) 8   (iii) 4

2. (i)   5 0 1 8 3 7 4 3 8 9 0 8 4
   (ii) 5 0 0 0 1 1 9 0 0 1 2 8 3
   (iii) 5 0 1 0 1 4 2 3 0 2 6 2 2
   (iv) 5 0 1 0 2 2 8 0 1 2 2 0 9

   (In each case there are many
      possible answers.)

Index